Collins

Edexcel GCSE Revision

Maths

Higher

Maths

Higher

Edexcel GCSE

Revision Guide

Linda Couchman
and Rebecca Evans

Contents

Contents

Contents

Contents

S Statistics   **P** Probability   **R** Ratio, Proportion and Rates of Change

Review Questions

Recap of KS3 Key Concepts

1 Write down two million and five as a number.   [1]

2 Work out 15% of 300kg.   [1]

3 Write 60 as the product of prime factors.   [1]

4 Find **a)** the HCF and **b)** the LCM of 36 and 90.   [2]

5 $6 + 4 \div 2 =$ _____   [1]

6 $(-2)^3 =$ _____   [1]

7 $(2^3)^2 =$ _____   [1]

8 Simplify: $5 + 2(m - 1)$ [1]

9 Simplify: $y^2 + y^2 + y^2$ [1]

10 **a)** Write the next two terms in the number sequence: 5, 8, 11, 14, __, __   [1]

    b) Is 140 a term in the sequence? Explain your answer.   [2]

11 $0.3 \times 0.2 =$ _____   [1]

12 A regular polygon has an interior angle of 140°. How many sides does the polygon have?   [1]

13 Work out $4.152 \div 1.2$ [1]

14 If $A = (2, 3)$ and $B = (7, 11)$, work out the coordinates of the midpoint of the line AB.   [2]

15 $3 \times 3 \times 3 =$ _____   [1]

16 Lorna left her home at 11.00am and went for a 20km run. She arrived back home at 1.00pm.

    Work out Lorna's average running speed.   [1]

17 The cross-section of a prism has an area of 25cm². The length of the prism is 5cm.

    Work out the volume of the prism.   [1]

18 The probability of it raining is 0.6

    Work out the probability of it **not** raining.   [1]

19 Write $\frac{7}{8}$ as a decimal.   [1]

20 Write down 0.78 as a fraction in its simplest form.   [2]

21 Find a fraction that lies between $\frac{2}{3}$ and $\frac{3}{4}$.   [2]

22 Pat threw three darts. The lowest score was 10. The range was 10. The mean was 15.

Work out the score for each dart. [3]

23 Write down the number that is one less than one million. [1]

24 Share £56 in the ratio 3 : 5 [2]

25 Put the following numbers in order, smallest first: 0.4, 0.04, 0.39, 0.394 [1]

26 Calculate a value for m if:

a) $m - 5 = -1$ [1]

b) $m \div 4 = 12$ [1]

c) $2m + 13 = 16$ [1]

27 A square has a perimeter of 24cm. Work out its area. [1]

28 Work out $1616 \div 4$. [1]

29 Temi spends £7.25. She pays with a £20 note.

How much change does she receive? [1]

30 Simplify $10a - 6y + 3a - 4y$ [1]

31 Increase £60 by 35%. [1]

32 A circle has a radius of 10cm. Taking π as 3.14, work out the area of the circle. [2]

33 How many cubes of side length 2cm will fit inside a hollow cube of side length 4cm? [3]

34 Write down the square root of 225. [1]

35 $\frac{3}{4}$ of a number is 63. Write down the number. [2]

36 Work out the value of m if $2 - m = 5$ [1]

37 Give the coordinates of the point where the graph of $x + y = 5$ crosses the x-axis. [2]

38 The two short sides of a right-angled triangle are 5cm and 12cm.

Work out the length of the longest side. [1]

39 If $a = 3$ and $b = 4$, work out the value of $2a^2 + 3b$ [2]

40 Find the median of 1, 4, 6, 13, 21, 10, 3 and 7. [1]

Total Marks _____ / 57

Order and Value

You must be able to:

- Order and compare positive and negative numbers
- Carry out calculations using positive and negative integers and decimals
- Understand and use standard form notation.

Negative Numbers

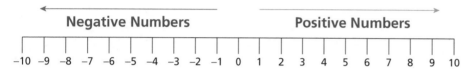

Negative Numbers   **Positive Numbers**

$$-10 \ -9 \ -8 \ -7 \ -6 \ -5 \ -4 \ -3 \ -2 \ -1 \ 0 \ 1 \ 2 \ 3 \ 4 \ 5 \ 6 \ 7 \ 8 \ 9 \ 10$$

- On a number line, positive numbers (+) are to the right of zero and negative numbers (–) are to the left of zero.
- Numbers on the left are smaller than numbers on the right.
- –4 is less than –2 and can be written $-4 < -2$
- –2 is greater than –4 and can be written $-2 > -4$
- A number line can help to visualise questions that involve negative numbers.

Key Point
A negative number is smaller than zero.

Put the following numbers in order, smallest to largest:

$-10.2 \quad -0.3 \quad -\frac{1}{5} \quad -6.4$

$-10.2 \quad -0.3 \quad -0.2 \quad -6.4$ ← Change all the numbers into decimals.

$-10.2 \quad -6.4 \quad -0.3 \quad -0.2$ ← Then rearrange by size.

Work out $2 - 7$

finish   start

$$-6 \ -5 \ -4 \ -3 \ -2 \ -1 \ 0 \ 1 \ 2 \ 3$$

$2 - 7 = -5$

Work out $-2 + (-2)$

finish   start

$$-6 \ -5 \ -4 \ -3 \ -2 \ -1 \ 0 \ 1 \ 2 \ 3$$

$-2 - 2 = -4$

Key Point
$-2 + (-2)$ and $-2 - (+2)$ both mean $-2 - 2 = -4$

Work out $-2 - (-2)$ ←  Two minus signs together become a plus.

$-2 + 2 = 0$

- When multiplying or dividing with two signs that are **different**, the answer is **negative**.
- When multiplying or dividing with two signs that are the **same**, the answer is **positive**.

$$-3 \times 6 = -18 \qquad 7 \times (-2) = -14 \qquad 24 \div (-6) = -4$$
$$-5 \times (-7) = 35 \qquad -100 \div (-5) = 20$$

Calculating with Decimals

- When **multiplying**, remove the decimal points and multiply the whole numbers.
- The number of decimal places in the answer is the same as the **total** number of decimal places in the original calculation.

Work out 0.6×0.2

$6 \times 2 = 12$ ← Remove the decimal points.

$\mathbf{0.6} \times \mathbf{0.2} = \mathbf{0.12}$ ← Two figures after the decimal points in the question, so two figures after the decimal point in the answer.

Work out 1.46×2.3

$146 \times 23 = 3358$

$1.46 \times 2.3 = 3.358$ ← Three figures after the decimal points on both sides of the equals sign.

- When **dividing**, change the problem so that you divide by a whole number.

Work out $27.632 \div 0.2$

$276.32 \div 2 = 138.16$ ← In order to divide by a whole number, multiply everything by 10.

Work out $36.14 \div 0.002$

$36\,140 \div 2 = 18\,070$ ← Multiply everything by 1000.

Standard Form

- **Standard form** is used to represent very small or very large numbers.
- A number in standard form is written in the form $A \times 10^{n}$, where $1 \leqslant A < 10$ and n is an integer.
- For numbers less than 1, n is negative.

Write the following numbers in standard form:

a) $723\,000$ $723\,000 = 7.23 \times 10^{5}$ ← $n = 5$ since the decimal point has to move 5 places to the right to go from 7.23 to 723 000

b) $0.000\,0063$ $0.000\,0063 = 6.3 \times 10^{-6}$ ← $n = -6$ since the decimal point has to move 6 places to the left to go from 6.3 to 0.000 0063

Types of Number

You must be able to:

- Use the ideas of prime numbers, factors (divisors), multiples, highest common factor, lowest common multiple and prime factors
- Show prime factor decomposition
- Use systematic listing strategies, including the product rule for counting.

Types of Number

- **Multiples** are found in the 'times table' of the number, e.g.
  Multiples of 4 = {4, 8, 12, 16 ...}
- A **factor** (divisor) is a number that will divide exactly into another number, e.g.
  Factors of 12 = {1, 2, 3, 4, 6, 12}
- A **prime number** has only two factors: itself and 1.
- **Square numbers** are the results of multiplying together two numbers that are the same. They are shown using a **power** of 2, e.g.

$1 \times 1 = 1$	$2 \times 2 = 4$	$3 \times 3 = 9$	$4 \times 4 = 16$
$1^2 = 1$	$2^2 = 4$	$3^2 = 9$	$4^2 = 16$

- A **square root** is the **inverse** (opposite) of a square, e.g.
  $6^2 = 36$ $\sqrt{36} = 6$
- **Cube numbers** are the results of multiplying together three numbers that are the same. They are shown using a power of 3, e.g.

$1 \times 1 \times 1 = 1$	$2 \times 2 \times 2 = 8$	$3 \times 3 \times 3 = 27$	$4 \times 4 \times 4 = 64$
$1^3 = 1$	$2^3 = 8$	$3^3 = 27$	$4^3 = 64$

- A **cube root** is the inverse of a cube, e.g.
  $5^3 = 125$ $\sqrt[3]{125} = 5$

Key Point

When listing factors, look for pairs of numbers, then you will not forget any.

Key Point

You are expected to know the squares and square roots up to $15 \times 15 = 225$ and that $\sqrt[3]{1} = 1$, $\sqrt[3]{8} = 2$, $\sqrt[3]{27} = 3$, $\sqrt[3]{64} = 4$, $\sqrt[3]{125} = 5$ and $\sqrt[3]{1000} = 10$

Prime Factors, LCM and HCF

- The **lowest common multiple (LCM)** of two numbers is the lowest **integer** that is a multiple of both numbers.

  Find the LCM of 12 and 18.

  Multiples of 12 = {12, 24, 36, 48 ...}

  Multiples of 18 = {18, 36, 54 ...}

  36 is the smallest number that is in both lists, so 36 is the LCM.

  Write out the multiples of 12 and 18 until you get a common value.

- The **highest common factor (HCF)** of two numbers is the largest integer that will divide exactly into both numbers.

  Find the HCF of 45 and 60.

  Factors of 45 = {1, 3, 5, 9, 15, 45}

  Factors of 60 = {1, 2, 3, 4, 5, 6, 10, 12, 15, 20, 30, 60}

  15 is the largest number that is in both lists, so 15 is the HCF.

  Write out the factors of 45 and 60 and look for the highest common value.

- Factors of a number that are also prime numbers are called **prime factors**.

Write 48 as a product of prime factors.

Method 1:
Prime Factor Decomposition

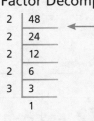

Method 2:
A Factor Tree

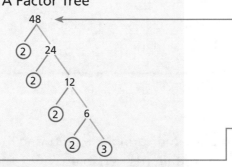

Keep splitting values into factors until the ends of all branches are prime numbers.

Keep dividing by prime numbers until you get to 1.

$$48 = 2 \times 2 \times 2 \times 2 \times 3 = 2^4 \times 3$$

Choices and Outcomes

- The **product rule** (multiplication rule) for counting is used to find the total number of **combinations** or **permutations** possible for a given scenario.
- It states: if there are A ways of doing Task 1 and B ways of doing Task 2, then there are $A \times B$ ways of doing both tasks.

In a restaurant there are three different starters (X, Y, Z) and six different main meals (1, 2, 3, 4, 5, 6) to choose from.

How many possible combinations are there when choosing a starter and a main meal?

Combinations are: X1, X2, X3, X4, X5, X6, Y1, Y2, Y3, Y4, Y5, Y6, Z1, Z2, Z3, Z4, Z5, Z6 (18 combinations)

Using the product rule:
3 (starters) × 6 (mains) = 18 (combinations)

How many different ways can you rearrange the letters PENCIL?

$6 \times 5 \times 4 \times 3 \times 2 \times 1 = 720$ ways

For the first letter, there is a choice of six, for the second letter, a choice of five, for the third letter, a choice of four, etc.

Key Point

A permutation is an arrangement of objects where the order is important. If order does not matter, it is a combination.

Key Words

multiple
factor
prime number
square number
power
square root
cube number
cube root
lowest common multiple
   (LCM)
integer
highest common factor
   (HCF)
prime factor
product rule
combination
permutation

Quick Test

1. Express 36 as a product of prime factors using any method.
2. Find **a)** the HCF and **b)** the LCM of 12 and 15.
3. Red buses leave the bus garage every 12 minutes. Blue buses leave the bus garage every 20 minutes. A red bus and a blue bus both leave the garage at 9am. At what time will a red bus and a blue bus next leave the garage together?

Basic Algebra

You must be able to:

- Use and understand algebraic notation and vocabulary
- Simplify and rearrange expressions, including those that involve algebraic fractions
- Expand and factorise expressions
- Solve linear equations.

Basic Algebra

- 'Like terms' are **terms** with the same **variable**, e.g. $3x$ and $4x$ are like terms because they both contain the variable x.
- To simplify an **expression** you must collect like terms.
- When moving a term from one side of an **equation** to the other, you must carry out the **inverse operation**.

> **Key Point**
>
> When simplifying expressions, remember to:
> - Use BIDMAS
> - Show your working.

Simplify $3x + 3y - 7x + y$

$4y - 4x$ ⟵

$3x - 7x = -4x$
$3y + y = 4y$

Simplify $9p^2 + 7p - qp + pq - p^2$ ⟵

$8p^2 + 7p$ ⟵

$qp = pq$

$9p^2 - p^2 = 8p^2$
$-qp + pq = 0$

- Substitution means replacing variables with numbers.

If $y = 4$ and $t = 6$, work out the value of $7y - 6t$

$7y - 6t = 7 \times 4 - 6 \times 6$

$\quad\quad\quad = 28 - 36$

$\quad\quad\quad = -8$

> **Key Point**
>
> An expression does not contain an = sign.

If $q = 5$, $r = 2$ and $z = -3$, work out the value of $rq + z^2$

$rq + z^2 = 2 \times 5 + (-3)^2$ ⟵

$\quad\quad\quad = 10 + 9$

$\quad\quad\quad = 19$

Use brackets as the minus sign is also squared.

> **Key Point**
>
> Always apply the rules:
> $- \times - = +$
> $+ \times + = +$
> $- \times + = -$
> $+ \times - = -$

- To expand or multiply out brackets, every term in the bracket is multiplied by the term outside the bracket.

Multiply out $5p(p - 2)$

$5p^2 - 10p$ ⟵

$5p \times p = 5p^2$
$5p \times (-2) = -10p$

Expand and simplify $4y(2y - 3) - 3y(y - 2)$

$8y^2 - 12y - 3y^2 + 6y$ ⟵

$= 5y^2 - 6y$

Note that $-3y \times -2 = +6y$

Factorisation

- **Factorisation** is the reverse of expanding brackets, i.e. you take out a common factor and put brackets into the expression.
- To factorise, you should look for common factors in every term.

Factorise $3x^2 - 6x$

$3x(x - 2)$

Factorise $3p^3 - 2p^2 + 8p$

$p(3p^2 - 2p + 8)$

> **Key Point**
>
> To factorise completely, always take out the highest common factor, e.g. 3 is the HCF of 3 and 6.

Linear Equations

Remember $x^2 = x \times x$

- Equations can be used to represent real-life problems.
- The equation should be rearranged to solve the problem.

Solve $5y - 4 = 3y + 10$

$5y - 3y = 10 + 4$

$2y = 14$

$y = 7$

Collect all the letter terms on one side.

> **Key Point**
>
> When moving a term from one side of an equation to the other, you must carry out the inverse operation.

Mary bought nine candles. She used a £3 gift voucher as part payment. The balance left to pay was £5.55

What was the cost of one candle (c)?

$9c - 3 = 5.55$

$9c = 8.55$

$c = \dfrac{8.55}{9}$

$c = £0.95$ or 95p

Use the information given to set up an equation.

Solve to find the cost of one candle.

Algebraic Fractions

- When solving equations involving **fractions**, take extra care when rearranging and make sure you carry out the inverse operations in the correct order.

Solve $\dfrac{3x - 3}{4} = 7$

$3x - 3 = 28$

$3x = 31$

$x = \dfrac{31}{3}$

Solve $\dfrac{5s}{4} + 3 = 18$

$\dfrac{5s}{4} = 15$

$5s = 60$

$s = 12$

Subtract 3 from both sides.

Multiply both sides by 4.

Divide both sides by 5.

Quick Test

1. Simplify $2y - 7 + 4y + 2$
2. Work out the value of $3p^3 - 7q$, when $p = -4$ and $q = -3$
3. Expand the following expression: $3t(4t - 1)$
4. Factorise completely $4r^3 - 2r^2$
5. Solve the equation $5(t + 4) = 3(6 - t)$

Factorisation and Formulae

You must be able to:

- Expand products of two or more binomials
- Factorise a quadratic expression
- Understand and use formulae
- Rearrange and change the subject of a formula.

Binomial Expansion

- A **binomial** is an **expression** that contains two terms, e.g. $3y^2 + 12$, $4xy - x^2$ or $5x^3 + 3x^2$
- The product of binomials is obtained when two or more binomials are multiplied together, e.g. $(2r + 7)(3r - 6)$
- To **expand** (or multiply out) the brackets, every term in the first set of brackets must be multiplied by every term in each of the other sets of brackets.

Expand $(2y + 4)(3y - 2)$

×	$2y$	$+4$
$3y$	$6y^2$	$+12y$
-2	$-4y$	-8

$6y^2 + 12y - 4y - 8$
$= 6y^2 + 8y - 8$ ⟵ Simplify by collecting like terms.

Expand $(4z + 3)(2z - 1)(4 - z)$

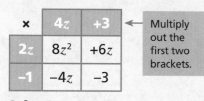 Multiply out the first two brackets.

×	$4z$	$+3$
$2z$	$8z^2$	$+6z$
-1	$-4z$	-3

$8z^2 + 6z - 4z - 3$
$= 8z^2 + 2z - 3$

×	$8z^2$	$+2z$	-3
4	$32z^2$	$+8z$	-12
$-z$	$-8z^3$	$-2z^2$	$+3z$

⟵ Multiply the product of the first two brackets with the final bracket.

$32z^2 + 8z - 12 - 8z^3 - 2z^2 + 3z$
$= -8z^3 + 30z^2 + 11z - 12$ ⟵ Write in descending powers of z.

Quadratic Factorisation

- An expression that contains a squared term is called **quadratic**.
- Some quadratic expressions can be written as a product of two binomial expressions.
- When written in factorised form, the new expression is equivalent to the original quadratic.
- To factorise a quadratic, you can use a table as in the expansion examples above:
  1. Complete the table with as much information as you can.
  2. Work out the missing terms (it may take more than one attempt to find the correct pair of numbers).
  3. Write out the factorised expression.

- If the **coefficient** of x^2 is **not** 1, more care must be taken.

Factorise the expression
$x^2 + 4x + 3$

×	x	$+1$
x	x^2	$+x$
$+3$	$+3x$	$+3$

The missing terms need to have a product of +3 and a sum of +4, i.e. 1 and 3.

$(x + 1)(x + 3)$

Write the expression as a product: the **first row** gives you the **first bracket** and the first column gives you the **second bracket**.

Factorise the expression
$2y^2 - 3y - 20$

×	$2y$	$+5$
y	$2y^2$	$+5y$
-4	$-8y$	-20

$(2y + 5)(y - 4)$

+5 and −4 is the pair that gives −3y.

Note that the coefficient of y^2 is not 1.

Possible pairs of numbers are −20 and 1; 20 and −1; 10 and −2; −10 and 2; −4 and 5; −5 and 4.

Changing the Subject of a Formula

- A **formula** is a way of describing a rule or fact.
- A formula is written as an algebraic equation.
- The **subject** of a formula appears once on the left-hand side.
- To change the subject, a formula must be rearranged using **inverse operations**.

This formula can be used to change temperature in degrees Fahrenheit to temperature in degrees Celsius: $C = \frac{5}{9}(F - 32)$

In Iceland, the lowest recorded temperature on a certain day is −20°C. What is this temperature in degrees Fahrenheit?

$$-20 = \frac{5}{9}(F - 32)$$
$$-180 = 5(F - 32)$$
$$-36 = F - 32$$
$$F = -4°F$$

The formula must be rearranged to find the value of F.

The answer is −4 degrees Fahrenheit.

Make r the subject of the formula $P = 3(r - 1)$

$$\frac{P}{3} = r - 1$$
$$r = \frac{P}{3} + 1$$

The formula for calculating the area of a circle is $A = \pi r^2$
Make r the subject.

$$\frac{A}{\pi} = r^2$$
$$r = \sqrt{\frac{A}{\pi}}$$

π can be treated as a numerical term.

Only the positive root is needed as r is a length.

Collect all the terms containing q on one side.

Make q the subject of the formula $p + q = pq - 5$

$$p + 5 = pq - q$$
$$p + 5 = q(p - 1)$$
$$q = \frac{p + 5}{p - 1}$$

Take q out as a factor.

> **Key Point**
>
> When rearranging formulae remember to use inverse operations. Finish by writing the formula out with the new subject on the left-hand side.

> **Key Words**
>
> binomial
> expression
> expand
> quadratic
> coefficient
> formula
> subject
> inverse operation

Quick Test

1. $T = 30w + 20$. Work out the value of w when $T = 290$
2. Expand and simplify $(y + 4)(y - 2)$
3. Factorise $2q^2 + 7q + 3$
4. Make y the subject of the formula $\frac{x + y}{3} = 2(y - 1)$

Ratio and Proportion

You must be able to:

- Apply ratio to real contexts and problems, including best buys
- Divide quantities into given ratios
- Use direct proportion
- Understand graphs that illustrate direct and inverse proportion.

Best Buys

- Working out the value of one part allows you to make like-for-like comparisons.

> Teabags are sold in two different sized packs.
> Pack A holds 80 teabags and costs £1.80
> Pack B holds 200 teabags and costs £3.80
> Which is the best buy?
>
>
>
> Pack A   180 ÷ 80 = 2.25p
> Pack B   380 ÷ 200 = 1.90p
>
> Pack B is the best buy (the cheapest price per teabag).

Key Point

If a ratio contains mixed units, change all values to the same unit before simplifying.

Work out the cost of one teabag in each pack.

Ratio

- You can **simplify** a **ratio**. This is like cancelling down a fraction.
- The unitary form of a ratio is $1 : n$

> Write down the ratio 5 : 9 in its unitary form.
>
> 5 : 9 = 1 : 1.8

Divide both sides by 5.

- To divide a quantity into a given ratio, you must first work out the value of one part.

> Sarah, John and James share £300 in the ratio 3 : 4 : 5
> How much money does each person receive?
>
> 3 + 4 + 5 = 12 parts in total, so £300 = 12 parts
> 300 ÷ 12 = £25
>   3 : 4 : 5 = (3 × £25) : (4 × £25) : (5 × £25)
>
> Sarah : John : James = £75 : £100 : £125

Divide by 12 to find the value of one part.

Always add up the values in your answer to check they equal the starting amount.

> An apple crumble recipe for four people uses 80g of plain flour.
> How much flour is needed for six people?
>
> 1 person uses 80 ÷ 4 = 20g
> 6 people use 6 × 20g = 120g

Divide by 4 to find the value of one part.

Direct and Inverse Proportion

- In a graph, **direct proportion** can be represented by a straight line that passes through the origin (0, 0).

This table gives information about the journey of a car during a half-hour time period.

Distance (miles)	0	6	12	18
Time (mins)	0	10	20	30

a) Plot the graph of distance against time.

b) Is the distance directly proportional to time? Give a reason for your answer.

c) At what speed was the car travelling in mph?

a)

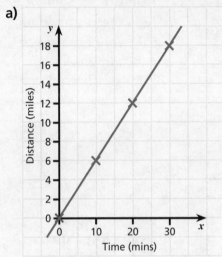

b) Distance is directly proportional to time, because it produces a straight line graph that passes through the origin.

c) Speed = $\dfrac{\text{Distance}}{\text{Time}}$

$= \dfrac{18 \text{ miles}}{30 \text{ minutes}}$

$= \dfrac{36 \text{ miles}}{60 \text{ minutes}}$

$= 36 \text{mph}$

This is the gradient.

- Quantities are in direct **proportion** if their ratio remains the same as they are increased or decreased.
- Direct proportion uses the symbol α.
- $y \, \alpha \, x$ means y is directly proportional to x or $y = kx$.
- k is called the **constant of proportionality**.
- Quantities can also be in **inverse proportion**. This is shown by $y \, \alpha \, \frac{1}{x}$ or $y = \frac{k}{x}$
- For example, for a journey of fixed distance (D), the time taken (T) is inversely proportional to the speed of travel (S), i.e. $T \, \alpha \, \frac{1}{s}$

Key Point

If quantities are in inverse proportion, as x increases, y decreases. The graph produced is a hyperbola (curve).

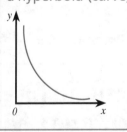

Quick Test

1. £180 is shared in the ratio 2 : 3
   What is the value of the largest share?
2. Simplify 3 hours : 45 minutes
3. a) Cat food comes in two different sized boxes:
   Box A has 12 sachets and costs £3.80
   Box B has 45 sachets and costs £14.00
   Which is the best buy? Explain your answer clearly.
   b) Tiggles the cat eats three sachets a day. How many of the best buy boxes have to be bought to feed Tiggles in June?

Key Words

simplify
ratio
direct proportion
proportion
constant of
    proportionality
inverse proportion

Variation and Compound Measures

You must be able to:

- Use compound units and solve problems involving compound measures
- Calculate compound interest
- Use iterative processes.

Compound Measures

- A **compound measure** is a measure that involves two or more other measures.
- **Speed** is a compound measure – it involves measures of distance and time.

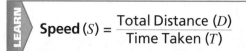

$$\text{Speed}\,(S) = \frac{\text{Total Distance}\,(D)}{\text{Time Taken}\,(T)} \qquad D = S \times T \qquad T = \frac{D}{S}$$

A snail crawls a distance of 30cm in 3 minutes 20 seconds. Work out its speed in m/s.

30cm = 0.3m ← Change 30cm into metres by dividing by 100 (100cm = 1m).

3 minutes 20 seconds = 200 seconds ←

Speed = Distance ÷ Time

    = 0.3 ÷ 200   Change the total time into seconds.

    = 0.0015m/s

- **Density** and **pressure** are compound measures.

$$\text{Density}\,(D) = \frac{\text{Mass}\,(M)}{\text{Volume}\,(V)} \qquad M = D \times V \qquad V = \frac{M}{D}$$

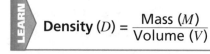

$$\text{Pressure}\,(P) = \frac{\text{Force}\,(F)}{\text{Area}\,(A)} \qquad F = P \times A \qquad A = \frac{F}{P}$$

Compound Interest and Repeated Percentage Change

- **Compound interest** is calculated based on the total amount of money invested, plus any interest previously earned.
- This formula can be used to calculate how money invested grows with time:

$$\text{Final Amount}\,(A) = \text{Original Amount} \times \left(1 + \frac{\text{Rate}}{100}\right)^{\text{time}}$$

- To calculate **depreciation**, the plus sign in the formula is changed to a minus sign.

Key Point

Units for speed include kilometres per hour (km/h) or metres per second (m/s).

Key Point

Always check that the units in the answer are the same as the units required in the question.

Key Point

Units for density include kg/m³ or g/cm³.

Pressure is expressed in newtons (N) per square metre (m²). A force of 1N applied to 1m² is called 1 pascal (Pa).

Key Point

Simple interest is when the interest is paid out, so each year starts with the same amount.

£3000 is invested at 2% compound interest per annum.
Calculate how much money there will be after four years.
Give your answer to the nearest pound.

$$A = 3000 \times \left(1 + \frac{2}{100}\right)^4$$
$$= 3000 \times (1.02)^4$$
$$= 3000 \times 1.0824$$
$$= £3247 \text{ (to the nearest £)}$$

Use the formula:
$A = \text{Original Amount} \times \left(1 + \frac{Rate}{100}\right)^{time}$

Compound interest questions only appear on a calculator paper.

The value of a new car is £8000.
The car depreciates in value by 10% each year.
Work out the car's value after five years.

$$A = 8000 \times \left(1 - \frac{10}{100}\right)^5$$
$$= 8000 \times (0.9)^5$$
$$= 8000 \times 0.5905$$
$$= £4723.92$$

Use the formula:
$A = \text{Original Amount} \times \left(1 - \frac{Rate}{100}\right)^{time}$

Trial and Improvement

- You can use trial and improvement (repetition) to find unknown values in a formula.

£600 is invested at r% compound interest.
After five years the value of the money has increased to £695.56
Using trial and improvement, find the value of r.

$$600 \times \left(1 + \frac{r}{100}\right)^5 = £695.56$$

Try $r = 6$ $600 \times (1.06)^5 = £802.94$ Too big.
Try $r = 4$ $600 \times (1.04)^5 = £729.99$ Slightly too big.
Try $r = 3$ $600 \times (1.03)^5 = £695.56$ Correct, so $r = 3\%$.
   $r = 3\%$

1. An aircraft travels 134 miles in 20 minutes. What is the aircraft's speed in mph?
2. Work out the compound interest on £1200 invested at 1.4% per annum for three years.
3. A new plasma television is worth £429. If you want to sell it two years later, it will be worth £267.74. Use trial and improvement to find the rate of depreciation.

compound measure
speed
density
pressure
compound interest
depreciation

Angles and Shapes 1

You must be able to:

- Recognise relationships between angles
- Use the properties of angles to work out unknown angles
- Recognise different types of triangle
- Understand and use the properties of special types of quadrilaterals.

Angle Facts

- There are three types of angle:
  - **acute:** less than 90°
  - **obtuse:** between 90° and 180°
  - **reflex:** between 180° and 360°.
- Angles on a straight line add up to 180°.
- Angles around a point add up to 360°.
- Vertically opposite angles are equal.

Angles in Parallel Lines

- Parallel lines never meet. The lines are always the same distance apart.
- **Alternate** angles are equal.
- **Corresponding** angles are equal.
- Co-interior or **allied** angles add up to 180°.

Work out the sizes of angles a, b, c and d.
Give reasons for your answers.

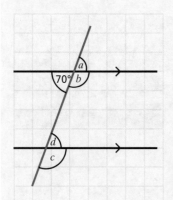

$a = 70°$ (vertically opposite angles are equal)

$b = 110°$ (angles on a straight line add up to 180°, so $b = 180° - 70°$)

$c = 110°$ (corresponding to b; corresponding angles are equal)

$d = 70°$ (corresponding to a; corresponding angles are equal)

Alternate Angles

Corresponding Angles

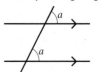

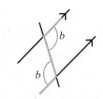

Allied Angles

$c + d = 180°$

> ### Key Point
>
> Examiners will **not** accept terms like 'Z angles' or 'F angles'. Always use correct terminology when giving reasons.

Triangles

- Angles in a triangle add up to 180°.
- There are several types of triangle:
  - **equilateral:** three equal sides and three equal angles of 60°
  - **isosceles:** two equal sides and two equal angles (opposite the equal sides)
  - **scalene:** no sides or angles are equal
  - **right-angled:** one 90° angle.

ABC is an isosceles triangle and HE is parallel to GD.
BAF is a straight line. Angle $FAE = 81°$

Calculate **a)** angle ABC and **b)** angle ACB.
Give reasons for your answers.

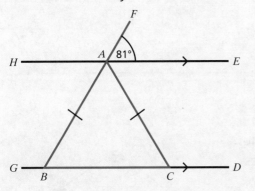

There are several different ways of solving this question.

a) Angle $HAB = 81°$ (vertically opposite FAE), so angle $ABC = 81°$ (alternate angle to HAB)

b) Angle $ACB = 81°$ (angle ABC = angle ACB; base angles of an isosceles triangle are equal.)

Special Quadrilaterals

- The interior angles in a quadrilateral add up to 360°.
- You need to know the properties of these special quadrilaterals:

	Sides	Angles	Lines of Symmetry	Rotational Symmetry	Diagonals
parallelogram	opposite sides are equal and parallel	diagonally opposite angles are equal	none	order 2	diagonals bisect each other
rhombus	all sides are equal and opposite sides are parallel	opposite angles are equal	two	order 2	diagonals bisect each other at 90°
kite	two pairs of adjacent sides are equal	one pair of opposite angles is equal	one	none	diagonals cross at 90°
trapezium	one pair of opposite sides is parallel		none (an isosceles trapezium has one)	none	

Quick Test

1. Name all the quadrilaterals that can be drawn with lines of lengths:
   a) 4cm, 7cm, 4cm, 7cm b) 6cm, 6cm, 6cm, 6cm
2. $EFGH$ is a trapezium with EH parallel to FG.
   FE and GH are produced (made longer) to meet at J.
   Angle $EHF = 62°$, angle $EFH = 25°$ and angle $JGF = 77°$.
   Calculate the size of angle EJH.

Key Words

acute
obtuse
reflex
alternate
corresponding
allied
equilateral
isosceles
scalene
right-angled
parallelogram
rhombus
kite
trapezium

Angles and Shapes 2

You must be able to:

- Work out angles in a polygon
- Answer questions on regular polygons
- Understand and use bearings.

Angles in a Polygon

- A **polygon** is a closed shape with at least three straight sides.
- **Regular** polygons are shapes where all the sides and angles are equal.
- **Irregular** polygons are shapes where some or all of the sides and angles are different.
- For all polygons:
  - at any **vertex** (corner): **interior** angle + **exterior** angle = 180°
  - sum of all exterior angles = 360°
- To work out the sum of the interior angles in a polygon, you can split it into triangles from one vertex.
- For example, a pentagon is divided into three triangles, so the sum of the interior angles is 3 × 180° = 540°
- The sum of the interior angles for any polygon can be calculated using the formula:

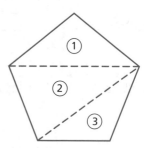

Pentagon | Exterior Angles | Interior Angles

LEARN

Sum of Interior Angles = $(n - 2) \times 180°$
where n = number of sides

Work out the sum of the interior angles of a decagon (10 sides).

Sum = (10 − 2) × 180° ← Use the formula: Sum = $(n - 2) \times 180°$
= 8 × 180°
= 1440°

Regular Polygons

- In regular polygons:

Number of Sides (n) × Exterior Angle = 360°
So, Exterior Angle = 360° ÷ n

Work out the size of the interior angles in a regular hexagon (6 sides).

Exterior angle = 360° ÷ 6 = 60° ← Use the formula: Exterior Angle = 360° ÷ n
Interior angle + 60° = 180° ← Interior Angle + Exterior Angle = 180°
Interior angle = 180° − 60°
= 120°

A regular polygon has an interior angle of 156°.

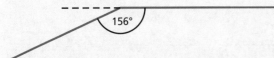

Work out the number of sides that the polygon has.

Exterior angle = 180° − interior angle
= 180° − 156°
= 24°
Number of sides = 360° ÷ 24°
= 15

Scale Drawings and Bearings

- **Bearings** are always measured in a clockwise direction from north (000°).
- Bearings always have three figures.

A ship sails from Mevagissey on a bearing of 130° for 22km.

a) Draw an accurate diagram to show this information and state the scale you have used.

b) What bearing would take the ship back to the harbour?

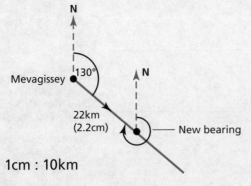

1cm : 10km

New bearing to return to harbour = 310°

Quick Test

1. For a regular icosagon (20 sides), work out **a)** the sum of the interior angles and **b)** the size of one interior angle.
2. A regular polygon has an interior angle of 150°.
   How many sides does the polygon have?
3. Two yachts leave port at the same time.
   Yacht A sails on a bearing of 040° for 35km.
   Yacht B sails on a bearing of 120° for 60km.
   Using a scale of 1cm : 10km, draw the route taken by both yachts.
   What is the bearing of yacht B from yacht A?

Key Words

polygon
regular
irregular
vertex
interior
exterior
bearing

Fractions

You must be able to:

- Add, subtract, multiply and divide fractions
- Change terminating and recurring decimals into corresponding fractions and vice versa
- Express one quantity as a fraction of another.

Adding, Subtracting and Calculating with Fractions

- When the **numerator** and **denominator** of a fraction are both multiplied or divided by the same number, the result is an **equivalent** fraction.
- To add or subtract fractions, you must first convert them to equivalent fractions with the same denominator.

$$5\frac{3}{5} + 1\frac{2}{7}$$

$$= 6\frac{21}{35} + \frac{10}{35}$$

$$= 6\frac{31}{35}$$

$$4\frac{1}{4} - 2\frac{3}{5}$$

$$= \frac{17}{4} - \frac{13}{5}$$

$$= \frac{85}{20} - \frac{52}{20}$$

$$= \frac{33}{20} = 1\frac{13}{20}$$

Multiplying and Dividing Fractions

- When multiplying fractions:
  - Convert mixed numbers to improper fractions
  - Cancel diagonally (or vertically) only
  - Then multiply the numerators together and the denominators together.
- When dividing fractions:
  - Invert the second fraction
  - Change the ÷ sign to a × sign
  - Then multiply.

$$6\frac{4}{5} \times 2\frac{1}{2}$$

$$= \frac{\overset{17}{\cancel{34}}}{\underset{1}{\cancel{5}}} \times \frac{\overset{1}{\cancel{5}}}{\underset{1}{\cancel{2}}} = \frac{17}{1} = 17$$

$$\frac{1}{4} \div \frac{7}{10}$$

$$= \frac{1}{\underset{2}{\cancel{4}}} \times \frac{\overset{5}{\cancel{10}}}{7} = \frac{5}{14}$$

> ### Key Point
>
> The top number in a fraction is called the **numerator**.
>
> The bottom number in a fraction is called the **denominator**.

Add whole numbers and then change fractions so that they have the same denominator.

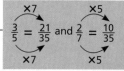

Convert mixed numbers into improper fractions and then change fractions so that they have the same denominator.

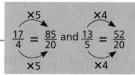

> ### Key Point
>
> A **mixed number** contains a whole number and a fraction.
>
> An **improper fraction** has a numerator larger than the denominator.

Cancel down and then multiply. If the answer is an improper fraction, convert it back to a mixed number.

Rational Numbers, Reciprocals, and Terminating and Recurring Decimals

- **Rational numbers** are numbers that **can be written exactly** as a fraction or decimal, e.g. $\frac{1}{4} = 0.25$
- π is an **irrational number** as it would continue forever as a decimal $3.141592\ldots$
- The **reciprocal** of a number (n) is 1 divided by the number, i.e. $\frac{1}{n}$
- A **terminating decimal** is a decimal with a finite number of digits (it ends), e.g. 0.75, 0.36
- A **recurring** decimal has a digit or group of digits that repeat indefinitely. It is shown by putting a dot over the recurring digit(s), e.g.

$0.1\dot{6} = 0.166666\ldots$ $0.\dot{3}\dot{7} = 0.37373737\ldots$

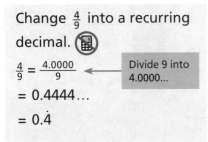

Change $\frac{4}{9}$ into a recurring decimal.

$\frac{4}{9} = \frac{4.0000}{9}$ ← Divide 9 into 4.0000...

$= 0.4444\ldots$

$= 0.\dot{4}$

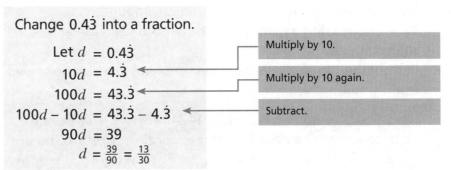

Change $0.4\dot{3}$ into a fraction.

Let $d = 0.4\dot{3}$

$10d = 4.\dot{3}$ ← Multiply by 10.

$100d = 43.\dot{3}$ ← Multiply by 10 again.

$100d - 10d = 43.\dot{3} - 4.\dot{3}$ ← Subtract.

$90d = 39$

$d = \frac{39}{90} = \frac{13}{30}$

One Quantity as a Fraction of Another

Express 20 minutes as a fraction of 3 hours 20 minutes.

3 hours 20 minutes = 200 minutes ← Convert to minutes.

20 minutes as a fraction of 200 minutes $= \frac{20}{200} = \frac{1}{10}$ ← Cancel down.

A school has 712 students.
$\frac{5}{8}$ of the students travel to school by bus and the rest walk. How many students walk to school?

Method 1 — Work out $\frac{1}{8}$ of 712 and then multiply by 5 to get $\frac{5}{8}$

$712 \div 8 = 89$

$89 \times 5 = 445$ (number of students who take the bus to school)

Number who walk = $712 - 445$

$= 267$ students

Method 2

$1 - \frac{5}{8} = \frac{3}{8}$ ← Start by working out how many students walk to school as a fraction.

$\frac{3}{8} \times 712 = 267$ students

Quick Test

1. Which is larger: $\frac{3}{4} + \frac{1}{5}$ or $\frac{3}{4} \div \frac{1}{5}$?
   You must show your working.
2. a) Change $\frac{1}{9}$ to a decimal.
   b) Is the resulting decimal terminating or recurring?
3. Express 15cm as a fraction of 4m in its simplest form.

Key Words

numerator
denominator
equivalent
rational number
irrational number
reciprocal
terminating decimal
recurring

Percentages

You must be able to:

- Increase and decrease quantities by a percentage
- Express one quantity as a percentage of another
- Solve percentage problems using a multiplier
- Understand and use reverse percentages.

Increasing and Decreasing Quantities by a Percentage

- When **increasing** a quantity by a **percentage**, work out the increase and then **add** it to the original amount.
- When **decreasing** a quantity, work out the decrease and then **subtract**.
- Percentage increase and decrease problems can also be solved using a **multiplier**.

> **Key Point**
>
> Percentages to learn:
>
> $100\% = 1$ (the whole)
>
> $20\% = \frac{1}{5}$
>
> $33.3\% = \frac{1}{3}$
>
> $66.7\% = \frac{2}{3}$
>
> $12.5\% = \frac{1}{8}$

Last year, Dandelion Place had 14 240 visitors. This year, the attraction will be closed for three months for refurbishment. This means there will be a 15% decrease in the total number of visitors for the year.

How many visitors are expected this year?

Method 1:

$\frac{15}{100} \times 14\,240 = 2136$

Number of visitors

$= 14\,240 - 2136$

$= 12\,104$

Method 2:

10% of 14 240 = 1424

5% of 14 240 = 712

15% of 14 240 = 2136

Therefore, the number of visitors = 14 240 − 2136

$= 12\,104$

> $10\% \div 2 = 5\%$

> $10\% + 5\% = 15\%$

> Alternatively, using a multiplier: 15% decrease means there will be 85% of last year's visitors, so $0.85 \times 14\,240 = 12\,104$

Molly bought a new tumble dryer.
She visited three shops to compare prices:

A. Sean's Electricals
   Normal price: £250, Sale price: 5% off

B. Mandeep's Deals
   Tumble dryer £200, plus tax at 17.5%

C. Chet's Cut Price Store
   Normal price: £280, Sale price: $\frac{1}{7}$ off

a) Which shop offered the best deal?
b) How much did it sell the tumble dryer for?

   A. Sean's Electricals:
      $\frac{5}{100} \times £250 = £12.50$
      Tumble dryer costs £250 − £12.50 = £237.50

    B. Mandeep's Deals:

    $\frac{17.5}{100} \times £200 = £35$

    Tumble dryer costs £200 + £35 = £235.00

    C. Chet's Cut Price Store:

    $\frac{1}{7} \times £280 = £40$

    Tumble dryer costs £280 – £40 = £240.00

  a) Mandeep's Deals **b)** £235

> If working without a calculator, remember:
> 17.5% = 10% + 5% + 2.5%

Sanjeev's salary is £32 000 per year. His salary is increased by 6%. Work out his new salary.

100% + 6% = 106%

106% = 1.06

1.06 × £32 000 = £33 920

> This is the multiplier.

Expressing One Quantity as a Percentage of Another

- Express the relationship between the two quantities as a fraction and then multiply by 100 to convert to a percentage.

> **Key Point**
>
> Make sure all quantities are in the same units first.

In two separate maths tests, Vinay got 18 out of 30 and Derek got 24 out of 45.

Who got the greater percentage in their maths test?

Vinay: $\frac{18}{30} \times 100 = 60\%$

Derek: $\frac{24}{45} \times 100 = 53.33\%$

Vinay got the greater percentage.

Work out 16 minutes as a percentage of 4 hours. Give your answer to 3 decimal places.

$\frac{16}{240}$

$\frac{16}{240} \times 100 = 6.667\%$ (to 3 d.p.)

> 4 hours = 4 × 60 = 240 minutes

Reverse Percentages

- Reverse percentages involve working backwards from the final amount to find the original amount.

In a Thai restaurant, there were $27\frac{1}{2}$ dumplings left on a plate after 45% had been eaten. How many dumplings were on the plate at the start of the meal?

100% – 45% = 55%

1% = 27.5 ÷ 55

100% = 100 × 27.5 ÷ 55 = 50 dumplings

> $27\frac{1}{2}$ dumplings is 55% of the original quantity.

Quick Test

1. This year Curt grew 220 carrots. This is 20% fewer than last year. How many carrots did Curt grow last year?
2. Increase £60 by 17.5%.
3. Express 18cm as a percentage of 12m.

> **Key Words**
>
> percentage
> multiplier

Probability 1

You must be able to:

- Know and use words associated with probability
- Construct and use a probability scale
- Understand what mutually exclusive events are
- Calculate probabilities using experimental data
- Categorise data into sets and subsets using tables and other diagrams.

Theoretical and Experimental Probability

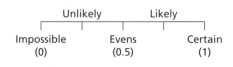

- The **probability** of an outcome occurring can be expressed using words or using a numerical scale from 0 to 1.
- **Relative frequencies** are probabilities based on experiments.
- **Random** means each possible outcome is equally likely.
- An event would be described as **biased** when outcomes are **not** equally likely.
- A sample space represents all possible outcomes from an event. This can be shown as a list or a diagram.

> **Key Point**
>
> Probabilities can be given as fractions, decimals or percentages.

Mutually Exclusive and Exhaustive Outcomes

- **Mutually exclusive** outcomes **cannot** happen at the same time.
- When two events are mutually exclusive P(A or B) = P(A) + P(B)
- A list of **exhaustive** events contains all possible outcomes.

> **Key Point**
>
> Probability is the chance of an event occurring. Probabilities can be based on theory or the results of an experiment. The sum of the probabilities of all possible outcomes is 1.

A spinner has three possible outcomes: A, B or C.
The probability that the spinner will land on A is 0.5
The probability that it will land on C is 0.3

a) Work out the probability that the spinner will land on B.

0.5 + 0.3 = 0.8
P(B) = 1 – 0.8 = 0.2

This is an exhaustive list.

The events are mutually exclusive.

b) Work out the probability that the spinner lands on A or B.

P(A or B) = 0.5 + 0.2 = 0.7

Expectation

- When you know the probability of an outcome, you can predict how many times you would expect that outcome to occur in a certain number of trials. This is called **expectation**.

> **Key Point**
>
> P(A') is the probability of A **not** occurring.
>
> P(A') = 1 – P(A)

Kelly rolls a six-sided dice 120 times and records her results.

Score	1	2	3	4	5	6
Frequency	19	29	14	18	20	20

a) Kelly throws the dice again. Estimate the probability that she will throw a 2.

Estimate of probability of a 2 is $\frac{29}{120}$

b) Is there enough evidence to suggest the dice is biased? Explain your answer.

If the dice is fair, you would expect Kelly to have rolled 20 of each number. This did not happen, so the dice may be biased. However, 120 is not a big sample. Kelly needs to roll the dice many more times. As she increases her sample size, her experiment will become more reliable and the results may come closer to the value of the theoretical probability.

Probability Diagrams

Julie thinks that in her class you are more likely to wear glasses if you are a boy. Based on the data below, is she correct?

	Boys	Girls	Total
Glasses	8	6	14
No Glasses	12	9	21
Total	20	15	35

P(of a boy wearing glasses) $= \frac{8}{20} = \frac{2}{5}$

P(of a girl wearing glasses) $= \frac{6}{15} = \frac{2}{5}$

Julie is incorrect as the probabilities are equal.

Students in a school study either French, Spanish, both or neither. There are 250 students. 75 students study both French and Spanish, 225 study French and 90 study Spanish.

Work out how many students do **not** study French or Spanish.

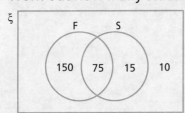

$250 - 150 - 75 - 15 = 10$

10 students do not study French or Spanish.

> **Key Point**
>
> To calculate the probabilities involving more than one event, use a **two-way table**, **Venn diagram** or tree diagram to represent all the possible outcomes.

Cancel down fractions.

The crossover of the circles represents students who study both subjects.

> **Key Words**
>
> probability
> relative frequency
> random
> biased
> mutually exclusive
> exhaustive
> expectation
> two-way table
> Venn diagram

> **Quick Test**
>
> 1. A spinner has four faces labelled 1, 2, 3 and 4. The probability that it will land on a 2 or 4 is given in the table.
>
> | Number | 1 | 2 | 3 | 4 |
> |---|---|---|---|---|
> | Probability | x | 0.3 | x | 0.5 |
>
> a) Work out the value of x.
> b) If you spin the spinner 150 times, how many times would you expect it to land on a 2?

Probability 2

You must be able to:

- Calculate the probability of independent and dependent combined events
- Calculate probabilities using a sample space diagram
- Calculate probabilities using a tree diagram
- Calculate and interpret conditional probabilities.

Addition Rules for Outcomes of Events

- When two outcomes are mutually exclusive, you can work out the probability of either outcome occurring by adding up the separate probabilities (see p.28).

Combined Events

- When calculating probabilities for combined events, you can draw a **sample space diagram**.

Two fair six-sided dice are thrown and their scores are added together.

a) Construct a sample space diagram to show all the possible outcomes.

b) Using your sample space diagram, work out P(8).

The sample space diagram shows there are 6 × 6 = 36 possible outcomes.

8 appears five times in the diagram out of 36 possible outcomes, therefore

$P(8) = \frac{5}{36}$

	Dice 1					
+	1	2	3	4	5	6
1	2	3	4	5	6	7
2	3	4	5	6	7	8
3	4	5	6	7	8	9
4	5	6	7	8	9	10
5	6	7	8	9	10	11
6	7	8	9	10	11	12

(Dice 2 labels the rows)

Tree Diagrams

- When dealing with successive events, you can use a **tree diagram**.

Two fair coins are tossed one after the other.

Write down the probability of each possible outcome.

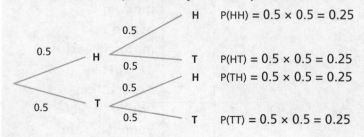

P(HH) = 0.5 × 0.5 = 0.25
P(HT) = 0.5 × 0.5 = 0.25
P(TH) = 0.5 × 0.5 = 0.25
P(TT) = 0.5 × 0.5 = 0.25

Each coin can give two possible outcomes: a Head (H) or a Tail (T).

Key Point

Always multiply along the branches of a probability tree.

Probability of two Heads = P(HH) = 0.25
Probability of two Tails = P(TT) = 0.25
Probability of a Head and a Tail = P(HT) + P(TH)
                                              = 0.25 + 0.25
                                              = 0.5 ←

0.25 + 0.25 + 0.5 = 1

Independent Events

- If A and B are independent events, one event does **not** depend on the other.
- If two events A and B are independent then P(A and B) = P(A) × P(B)

A fair two-sided coin is tossed and a fair six-sided dice is rolled.

What is the probability of getting a Head and rolling a 6?

P(H) × P(6) = $\frac{1}{2}$ × $\frac{1}{6}$ ←
              = $\frac{1}{12}$

The two events are independent, therefore P(A) × P(B)

Conditional Probability

- If the probability of an event occurring is dependent on the outcome of the previous event, it is called **conditional** probability.

Philippa has a box of chocolates that contains 7 milk chocolates, 4 dark chocolates and 3 white chocolates.

She takes two chocolates from the box without looking. What is the probability that both chocolates are the same?

Probability that both are the same = P(MM) + P(DD) + P(WW)
$\left(\frac{7}{14} \times \frac{6}{13}\right) + \left(\frac{4}{14} \times \frac{3}{13}\right) + \left(\frac{3}{14} \times \frac{2}{13}\right) = \frac{30}{91}$

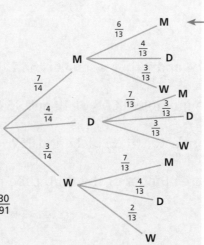

If one of the milk chocolates is taken, only 13 chocolates remain and 6 of them will be milk.

Quick Test

1. A fair three-sided spinner has one red section, one green section and one blue section all of equal size. The spinner is spun and a fair six-sided dice is thrown. What is the probability of getting a 5 on the dice with a red on the spinner?
2. A bag contains three red balls, four blue balls and two yellow balls. A ball is drawn from the bag and **not** replaced. A second ball is then drawn from the bag. What is the probability that both balls are the same colour?

 Key Words

sample space diagram
tree diagram
independent
conditional

Practice Questions

Order and Value

1. Write the following numbers in order, smallest to largest: 🖩

   -3 -3.38 $-\frac{1}{8}$ 3.3 [2]

2. Which is larger, $-4 \times -4 \times -4$ **or** $(2^3 - 3^3) - (-7)^2$? Show your working. 🖩   [3]

3. The average distance from the Moon to the Earth is 384 000 km.

   a) Write this distance in standard form. 🖩   [2]

   b) A spaceship is travelling from the Earth to the Moon.
   It has travelled 1.96×10^4 km.

   How many kilometres are left to travel?   [2]

> **Total Marks** _____ / 9

Types of Number

1. What is the largest multiple of 4 and 7 that is smaller than 105? 🖩   [1]

2. a) What is the lowest common multiple (LCM) of 12 and 15? 🖩   [1]

   b) What is the highest common factor (HCF) of 12 and 15?   [1]

3. Which is greater, the fifth square number or the third cube number?   [2]

4. 256 can be expressed as 2^n.

   What is the value of n?   [2]

5. Is 2^3 equal to 3^2?
   Give a reason for your answer. 🖩   [2]

6. Sausages come in packs of six. Bread rolls come in packs of four.

   What is the least number of packs of sausages and packs of bread rolls that Linda must buy, so that there is a roll for every sausage without any left over? 🖩   [2]

> **Total Marks** _____ / 11

Basic Algebra & Factorisation and Formulae

1   Simplify $5x - 2y + 4x + 6y$ 🖩   [2]

2   Simplify $5 - 3z + y - 5z + 7 + 3y$ 🖩   [2]

3   Simplify $3x^2 + 3x + x^2 + 4 - x$ 🖩   [2]

4   Work out the value of $3z^2 - 2q + 5$ when $z = -2$ and $q = -3$ 🖩   [1]

5   Solve $4(2b - 3) = 2$ 🖩   [2]

6   Solve $3(p + 2) = 2(p + 3)$ 🖩   [2]

7   Solve $\frac{5}{2}x - \frac{1}{3} = \frac{2}{3}x + \frac{1}{2}$ 🖩   [3]

8   Expand $6(x - 5y + 6)$ 🖩   [2]

9   Expand $6p - 4(q - 3)$ 🖩   [2]

10   Factorise completely $4xyz - 4xz$ 🖩   [2]

11   Expand and simplify $(w + 4)(w + 1)(w - 4)$ 🖩   [3]

12   Factorise $x^2 + 3x + 2$ 🖩   [2]

13   Write $3(2x - 5) + 4(x + 3) - 4x$ in the form $a(bx + c)$, where a, b and c are integers. 🖩   [3]

14   Rearrange the formula to make y the subject: $x = \dfrac{1 - 2y}{3 + 4y}$ 🖩   [3]

15   The formula for the area of a trapezium is $A = \frac{1}{2}(a + b)h$ 🖩

   a)   Rearrange the formula to make h the subject.   [2]

   b)   The area of a trapezium is 24cm². Work out the value of h if $a = 5$cm and $b = 7$cm.   [2]

16   Factorise $x^2 + 5x + 6$ 🖩   [2]

17 $V = \sqrt{u^2 - 10p}$ 🖩

   a)   Work out the value of V when $u = 10$ and $p = 5$.   [2]

   b)   Rearrange the formula to make u the subject.   [2]

Total Marks _____ / 41

Practice Questions

Ratio and Proportion

1 Simplify 5g : 10kg 🖩 [1]

2 The angles in a triangle are in the ratio 2 : 3 : 4

What is the size of the largest angle? 🖩 [2]

3 Six sticks of celery are needed to make celery soup for four people.

How many sticks of celery would be needed to make soup for 14 people? 🖩 [2]

4 It took six people four days to build a wall.

a) Working at the same rate, how long would it have taken eight people to build the wall? [2]

b) Working at the same rate, how many people would have been needed if the wall had to be completed in two days? [1]

Total Marks .... / 8

Variation and Compound Measures

1 A bar of lead has a volume of 400cm³ and a mass of 4.56kg.

Work out the density of the bar of lead in g/cm³. [2]

2 A rabbit runs 200 metres in 22 seconds.

What is the rabbit's average speed in m/s?
Give your answer to 2 decimal places. [2]

3 Khalid left his home at 10am and went for a 15km run.
He arrived back home at 1pm.

What was his average speed in km/h? [2]

4 Work out the compound interest earned on £4000 invested at 4% for four years. [3]

Total Marks .... / 9

Angles and Shapes 1 & 2

1 Work out the size of angles j, k, l and m, giving a reason for each answer. 🖩

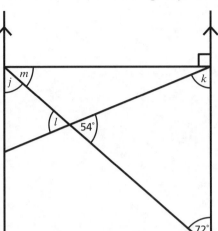

[4]

2 $ABCD$ is a parallelogram. 🖩

AB is parallel to CD and AD is parallel to BC. Angle $BAD = 110°$

Work out:

a) Angle DCB. [1]

b) Angle ABC. [1]

3 The angles in a quadrilateral are x, $2.5x$, $3x$ and $2.5x$ degrees.

Calculate the size of the largest angle. [2]

4 Work out the interior angle of a regular decagon. 🖩 [2]

5 A and B are two points.

If the bearing of B from A is 036°, what is the bearing of A from B? 🖩 [1]

6 A regular polygon has an exterior angle of 45°. 🖩

a) Work out how many sides the polygon has. [1]

b) What is the name of the polygon? [1]

Total Marks .... / 13

Fractions

1 A teacher took 32 books home to mark. She marked $\frac{1}{8}$ of them.

How many books does she still have to mark? [2]

2 On a farm, $\frac{1}{3}$ of the livestock is cows, $\frac{1}{6}$ is sheep, $\frac{1}{4}$ is chickens and the remainder is horses.

Express how many horses there are on the farm as a fraction. [2]

3 Write these fractions in ascending order:

$\frac{2}{3}$ $\frac{4}{12}$ $\frac{1}{6}$ $\frac{12}{24}$ [2]

4 **a)** Work out the area of a rectangle measuring $\frac{4}{5}$ of a metre by $\frac{2}{3}$ of a metre. [1]

b) What is the rectangle's perimeter? Give your answer as an improper fraction. [2]

5 Change $0.1\dot{8}$ to a fraction. [2]

Total Marks _____ / 11

Percentages

1 Zahra earns £24 000 per year.
The first £6000 is tax free. She pays 22% tax on her remaining salary.

How much tax does she pay? [2]

2 Express 46 seconds of 2 hours as a percentage. [1]

3 Whizzy Garage sells cars. It offers a discount of 20% off the retail price for cash purchases.
Pat pays £4800 cash for a car.

Calculate the retail price of the car before the cash discount. [2]

Total Marks _____ / 5

Probability 1 & 2

1 Two fair six-sided dice are thrown and the product of their scores is calculated. 🖩

   a) Draw a sample space diagram to represent all the possible outcomes.   [2]

   b) What is the probability that the product of the scores is greater than 12?   [1]

   c) What is the probability that the product of the scores is a square number?   [1]

2 There are 20 pens in a box: ten black, six blue and four red. 🖩
A pen is taken at random from the box.

   a) What is the probability that the pen is black?   [1]

   b) What is the probability that the pen is **not** black?   [1]

3 A three-sided spinner has sides labelled 1, 2 and 3.
The spinner is biased.
Thomas spins the spinner 500 times and records the outcomes in a table.

Score	1	2	3
Frequency	98	305	97

   a) Thomas spins the spinner again. Estimate the probability that it lands on a 2.   [1]

   b) He takes another 100 spins. Estimate how many times it will land on a 3.   [2]

4 The probability that it will rain (R) on Wednesday is 0.65
If it does rain on Wednesday, the probability that it
will **not** rain (R') on Thursday is 0.37
If it does **not** rain on Wednesday, the probability it will
rain on Thursday is 0.55

   a) Complete the probability tree diagram.   [3]

   b) Calculate the probability that it rains on one of the days.   [3]

5 Bhavna conducts a survey about the pets owned by her classmates.
There are 35 students in her class. Her results are: 25 students own cats,
15 students own dogs and 10 students own both. 🖩

   a) Draw a Venn diagram to represent the results.   [3]

   b) How many students do **not** own a cat or a dog?   [1]

Total Marks _____ / 19

You must be able to:

- Work out missing terms in sequences using term-to-term rules and position-to-term rules
- Recognise and use arithmetic and geometric sequences
- Work out the rule for a given pattern.

Patterns in Number

- A **sequence** is a series of shapes or numbers that follow a particular pattern or rule.
- A **term-to-term rule** links the next term in the sequence to the previous one.
- A **position-to-term rule**, also called the *n*th **term**, can be used to work out any term in the sequence.
- A **recursive** relationship is used to define further terms in a sequence, when one or more initial terms are given.

> Write down the next two terms in the following sequence.
>
> 7, 11, 15, 19, __, __
>
> The term-to-term rule is +4, so the next two terms are 23 and 27.
>
> This sequence can be expressed by the recursive relationship $U_{n+1} = U_n + 4$, $U_1 = 7$
>
> This states that the first term is 7.
> Therefore, U_5 (the fifth term) = $U_4 + 4 = 19 + 4 = 23$

General Rules from Given Patterns

> Here is a sequence of patterns made from matchsticks:
>
>
>
> pattern 1   pattern 2   pattern 3
>
> **a)** Draw the next pattern in the sequence.
>
>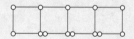
>
> **b)** Write down the sequence of numbers that represents the total number of matchsticks used in each pattern and work out an expression for the position-to-term rule.

The expression for the position-to-term rule is $3n + 1$

Key Point

For the first term in any sequence, $n = 1$

Pattern No.	1	2	3	4	*n*
No. of Matchsticks	4	7	10	13	$3n + 1$

The nth term of a number sequence is $5n + 2$.

Write down the first five terms of the sequence.

$(5 \times 1) + 2 = 7$
$(5 \times 2) + 2 = 12$
7, 12, 17, 22, 27

To work out the first term, substitute $n = 1$ into the expression.

To work out the second term, substitute $n = 2$ into the expression.

Continue until $n = 5$ to produce the first five terms of the sequence.

Number Sequences

- In an **arithmetic sequence** there is a common difference between consecutive terms, e.g.
  5, 8, 11, 14, 17 …

  The terms have a common difference of +3.

- In a **geometric sequence** each term is found by multiplying the previous term by a constant, e.g.
  20, 10, 5, 2.5, 1.25 …

  The constant (or ratio) is 0.5

The nth term of an arithmetic sequence is $4n - 1$.

a) Write down the term-to-term rule.

The sequence of numbers is 3, 7, 11, 15, 19 …
The term-to-term rule is +4.

Work out the first five terms.

b) Marnie thinks that 50 is a number in this sequence. Is Marnie correct? Explain your answer.

$4n - 1 = 50$, $n = 12.75$

n is not a whole number, so 50 is **not** in this sequence. Marnie is wrong.

Here is a geometric sequence: 4, 6, 9, __, 20.25

What is the missing term?

$\frac{6}{4} = \frac{9}{6} = 1.5$

Divide at least two given terms by the previous term to work out the ratio.

$9 \times 1.5 = 13.5$
The missing term is 13.5

Multiply by the ratio to find the missing term.

Quick Test

1. Here are the first five terms of a sequence: 16, 12, 8, 4, 0 … Write down the next two terms.
2. The nth term of a sequence is $5n - 7$. Work out the 1st term and the 10th term of this sequence.
3. In the sequence below the next pattern is formed by adding another layer of tiles around the previous pattern.

Work out how many tiles will be needed for the 6th pattern.

Key Words

sequence
term-to-term rule
position-to-term rule
nth term
recursive
arithmetic sequence
geometric sequence

Terms and Rules

You must be able to:

- Deduce and use expressions to calculate the nth term of a linear sequence
- Deduce and use expressions to calculate the nth term of a quadratic sequence
- Recognise and use sequences of triangular numbers, squares, cubes and other special sequences.

Finding the nth Term of a Linear Sequence

- A number sequence that increases or decreases by the same amount each time is called a **linear sequence**.
- To work out the expression for the nth term in a linear sequence, look for a pattern in the numbers.
- Using a function machine to represent a sequence of numbers can help.

The first five terms of a sequence are: 9, 12, 15, 18, 21 ...

What is the expression for the nth term of this sequence?

Input (n)	× 3 ($3n$)	Output
1	3	9
2	6	12
3	9	15
4	12	18
5	15	21
n	$3n$	$3n + 6$

The term-to-term rule is +3, so the expression for the nth term starts with $3n$.

The 'input' is the position of the term and the 'output' is the value of the term.

The difference between $3n$ and the output in each case is 6, so the expression for the nth term is **$3n + 6$**.

 Key Point

The zero term is the term that would come before the first term in a given sequence of numbers.

- The alternative method is to work out the **zero term**.

The first five terms of a sequence are: 20, 16, 12, 8, 4 ...

What is the expression for the nth term of this sequence?

Input	Output
0	zero term
1	20
2	16
3	12
4	8
5	4
n	nth term

The difference between terms is –4.

The zero term is (20 + 4 =) 24

The expression for the nth term is $-4n + 24$ or $24 - 4n$.

The nth term = 'the difference' × n + the zero term

Special Sequences

- It is important to be able to recognise special sequences of numbers:
  - square numbers: 1, 4, 9, 16, 25 ...
  - cube numbers: 1, 8, 27, 64, 125 ...
  - triangular numbers: 1, 3, 6, 10, 15 ...
  - the **Fibonacci sequence**: 1, 1, 2, 3, 5, 8, 13, 21 ...

> The nth term is n^2

> The nth term is n^3

> The nth term is $\frac{n}{2}(n + 1)$

> The recursive relationship can be written as $U_{n+2} = U_{n+1} + U_n$

Work out the expression for the nth term of the following sequence: 2, 9, 28, 65, 126 ...

The expression for the nth term is $n^3 + 1$

> The cube numbers produce the sequence 1, 8, 27, 64, 125 ... This sequence of numbers is one greater than the cube numbers.

Finding the nth Term of a Quadratic Sequence

- A **quadratic sequence** contains an n^2 term as part of the expression for the nth term.
- The general form of a quadratic sequence is $an^2 + bn + c$, where a, b and c are constants.
- To find the expression for the nth term of a quadratic sequence, look to see if you can spot the pattern.
- If this doesn't work, extend the difference backwards to get the zero term ($n = 0$).

Work out the expression for the nth term of the following sequence of numbers: 0, 3, 8, 15, 24, 35 ...
$n^2 - 1$

> The numbers are one less than the square numbers.

Work out the expression for the nth term of the following sequence of numbers: 4, 13, 26, 43 ...

n	0	1	2	3	4
c (nth Term)	−1	4	13	26	43
$a + b$ (First Difference)		5	9	13	17
$2a$ (Second Difference)			4	4	4

> Set up a difference table that extends backwards to get the term for $n = 0$.

$c = -1$
$2a = 4$, so $a = 2$ $a + b = 5$, so $b = 3$
The expression for the nth term is $2n^2 + 3n - 1$

> In the form $an^2 + bn + c$

Quick Test

1. a) Work out the expression for the nth term for the following sequence: 7, 10, 13, 16, 19 ...
    b) Work out the 50th term in this sequence.
2. Write down the next two terms in this quadratic sequence: −4, −1, 4, 11, 20, __, __
3. Write down an expression for the nth term for the following sequence of numbers: 0, 7, 26, 63, 124 ...

Key Words

linear sequence
zero term
Fibonacci sequence
quadratic sequence

Transformations

You must be able to:

- Identify, describe and construct transformations of shapes, including reflections, rotations, translations and enlargements
- Describe the changes achieved by combinations of transformations.

Transformations

- **Reflection:**
  - Each point on the image is the same distance from the mirror line as the corresponding point on the object
  - The object and the image are **congruent** (same size and shape)
  - To define a reflection on a coordinate grid, the equation of the mirror line should be stated.
- **Rotation** is described by stating the:
  - Direction rotated (clockwise or anticlockwise)
  - Angle of rotation (which is usually a multiple of 90° in the exam)
  - Centre of rotation (point about which the shape is rotated).

> **Key Point**
>
> There is no need to state clockwise or anticlockwise for a rotation of 180°.

a) Describe the transformation that maps triangle ABC onto triangle $A'B'C'$.

The transformation that maps ABC to $A'B'C'$ is a rotation of 90° anticlockwise (or 270° clockwise) about the origin (0, 0).

b) Describe the transformation that maps triangle $A'B'C'$ onto triangle $A''B''C''$.

The transformation that maps $A'B'C'$ to $A''B''C''$ is a rotation of 180° about the origin (0, 0).

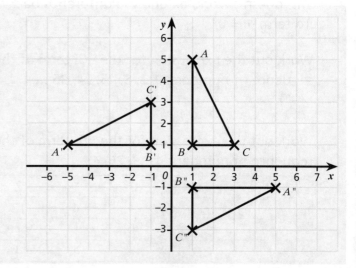

- **Translation:**
  - The shape does not rotate – it moves left or right and up or down – and stays the same size
  - The translation is represented by a **column vector** $\begin{pmatrix} x \\ y \end{pmatrix}$.

Describe the transformation that takes shape A to shape B.

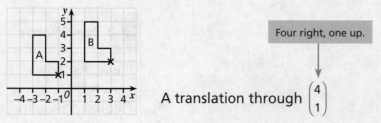

Four right, one up.

A translation through $\begin{pmatrix} 4 \\ 1 \end{pmatrix}$

> **Key Point**
>
> x represents the distance moved **horizontally**: **positive** means to the **right** and **negative** means to the **left**.
>
> y represents the distance moved **vertically**: **positive** means **up** and **negative** means **down**.

- Enlargement:
  - The shape of the object is not changed, only its size. The enlarged shape is **similar** to the original shape
  - The **scale factor** determines whether the object gets bigger (scale factor > 1) or smaller (scale factor < 1)
  - Scale factors can be negative. This results in:
    - A 180° rotation of the object
    - The image being on the opposite side of the centre of enlargement to the object
  - When describing the enlargement, state the scale factor and the centre of enlargement.

Enlarge triangle A by scale factor 2, centre of enlargement (1, 2). Label the transformed triangle B.

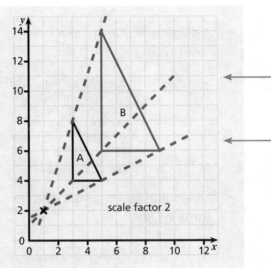

All construction lines must remain.

The side **lengths** of triangle B are **twice** the length of the corresponding sides of triangle A. However, the **area** of triangle B is **four times** bigger.

Combinations of Transformations

- It is possible for two or more transformations to be applied to a shape, one after the other.

Enlarge triangle C by scale factor $-\frac{1}{2}$ about centre of enlargement (0, 0) to form triangle D.

Transform triangle D by a reflection in the line $y = 1$ to form triangle E.

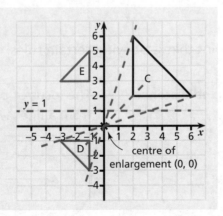

Quick Test

1. Describe the single transformation that takes:
   a) triangle A to triangle B
   b) triangle A to triangle C
   c) triangle A to triangle D.

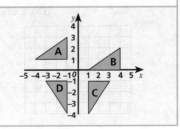

Constructions

You must be able to:

- Use a ruler and a pair of compasses to produce different constructions, including bisectors
- Describe a locus and solve problems involving loci
- Understand and construct plans and elevations of 3D shapes.

Constructions

- The **perpendicular bisector** of line AB.

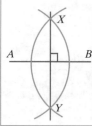

Open compasses to more than half AB. Put compass point on A. Draw arc. Put compass point on B. Draw arc. XY is the perpendicular bisector.

- A **perpendicular** from a given point to the line AB.

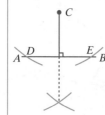

Put compass point on C. Draw arc. Keep radius the same. Put compass point on E. Draw arc. Put compass point on D. Draw arc. Join C to the point where the arcs cross.

- An **angle bisector**.

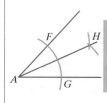

Put compass point on A. Draw arc FG. Put compass point on F. Draw arc. Put compass point on G. Draw arc. Join HA.

- An **equilateral triangle** and an **angle of 60°**.

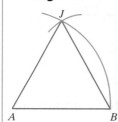

Open compasses to length AB. Put point on A. Draw arc. Put compass point on B. Draw arc. Join AJ and BJ. Angle $A = 60°$

> **Key Point**
>
> The perpendicular distance from a point to a line is the shortest distance to the line.

> **Key Point**
>
> To construct a perpendicular at a given point on a line:
>
> Put the compass point on that point.
>
> Draw two arcs on the line either side of that point.
>
> Then construct the perpendicular bisector of the two new points.

Defining a Locus

- A **locus** is the path taken by a point that is obeying certain rules.
- The plural of locus is **loci**.

- The locus of points that are a **fixed distance from a given point** A is a circle.

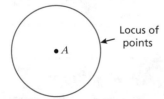

- The locus of points that are a **fixed distance from a line** AB.

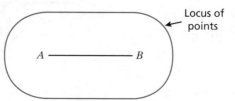

- The locus of points that are the **same distance from two lines** AB and AC. This is the angle bisector.

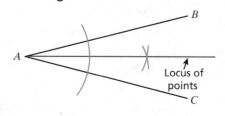

- The locus of points that are **equidistant**, or the **same distance, from two points** A and B. This is the perpendicular bisector.

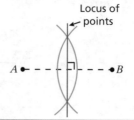

Loci Problems

A guard dog is tied to a post by a 4-metre long rope.

Accurately draw the locus of the points the dog can reach using a scale of 1cm : 1m

The solution would be a shaded circle of radius 4cm.
The dog could reach the circumference of the circle and all points within it.

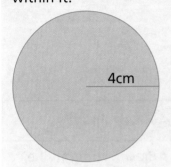

4cm   Not Drawn Accurately

Plans and Elevations

- The **plan view** of a 3D shape shows what it looks like from above – a bird's eye view.
- The side **elevation** is the view of a 3D shape from the side.
- The front elevation is the view from the front.

Here is a 3D shape made from centimetre cubes drawn on isometric paper.

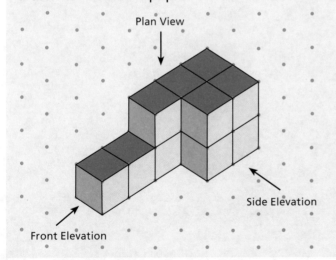

Plan View

Side Elevation

Front Elevation

On squared paper, draw:
a) the plan view

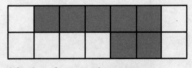

b) the front elevation

c) the side elevation.

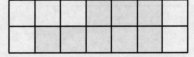

> **Key Words**

perpendicular
bisector
locus / loci
equidistant
plan view
elevation

Linear Graphs

You must be able to:

- Work with coordinates in all four quadrants
- Plot graphs of linear functions
- Work out the equation of a line through two given points or through one point with a given gradient
- Work out the gradient and y-intercept of a straight line in the form $y = mx + c$.

Drawing Linear Graphs from Points

- **Linear graphs** are straight-line graphs.
- The equation of a straight-line graph is usually given in the form $y = mx + c$, where m is the **gradient** of the line and c is the **intercept** of the y-axis.
- $y = mx + c$ is a function of x, where the input is the x-coordinate and the output is the y-coordinate.

> **Key Point**
>
> To draw a straight line, only two coordinates are needed.

Draw the graph of the equation $y = 2x + 5$

Use values of x from –3 to 3.

x	–3	0	3
y	–1	5	11

First, draw a table of values. Include a third value as a check.

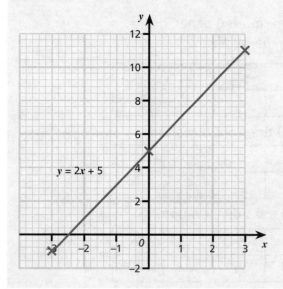

$y = 2x + 5$

Drawing Graphs by Cover-Up and Gradient–Intercept Methods

- Another method that can be used to draw a graph is the cover-up method.
- This can be used for equations in the form $ax + by = c$.

Draw the graph of the equation $2x + 3y = 6$

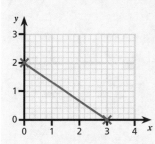

$3y = 6$

$y = 2$ The y-intercept is (0, 2).

$2x = 6$

$x = 3$ The x-intercept is (3, 0).

Cover up the x term and solve to find y.

Cover up the y term and solve to find x.

- The gradient–intercept method is used for equations in the form $y = mx + c$.

Draw the line with equation $y = 4x + 2$

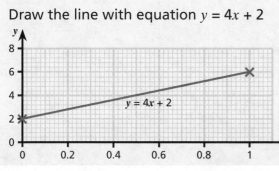

The y-intercept is (0, 2).

The gradient is 4.

Therefore, another coordinate on the line is (1, 6).

Find the value for y when $x = 0$.

m is the gradient in the equation $y = mx + c$

Gradient $= \frac{4}{1}$ so for every 1 across, go up 4.

Finding the Equation of a Line

- To find the equation of a straight line in the form $y = mx + c$, work out the gradient and y-intercept.

Work out the equation of the line that joins the points (1, 20) and (4, 5).

Gradient $= \frac{15}{-3}$

$= -5$

$y = mx + c$

$y = -5x + c$

$20 = (-5 \times 1) + c$

$c = 25$

The equation of the line is $y = -5x + 25$

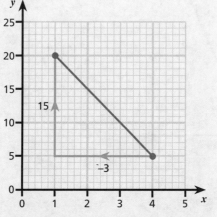

Key Point

Gradient $= \dfrac{\text{Change in } y}{\text{Change in } x}$

If the line slopes down from left to right, the gradient is negative.

To work out the value of c, substitute in point (1, 20) or (4, 5).

Quick Test

1. Draw the graph with equation $y = 4x + 1$
2. Write down the gradient and y-intercept of the line with equation $y = 5 - 2x$
3. Work out the equation of the line that joins the points (5, 7) and (3, 10).

Key Words

linear graph
gradient
intercept

Graphs of Quadratic Functions

You must be able to:

- Recognise, sketch and interpret graphs of quadratic functions
- Identify and interpret roots, intercepts and turning points of quadratic functions
- Work out roots using algebraic methods
- Work out turning points.

Plotting Quadratic Graphs

- A **quadratic equation** is an equation that contains an unknown term with a power of 2, e.g. x^2.
- You can use a table of values to draw **quadratic graphs**.

Draw the graph of the function $y = 2x^2 + 1$

x	−2	−1	0	1	2	3	4
y	9	3	1	3	9	19	33

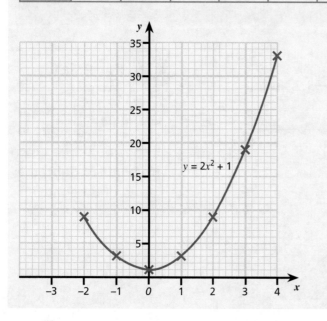

$y = 2x^2 + 1$

The Significant Points of a Quadratic Curve

- A sketch shows the shape and significant points on a graph, but is not an accurate drawing.
- To sketch a quadratic, work out the **roots**, the **intercept** and the **turning point**, i.e. the **maximum** or **minimum point**.
- The roots are found by solving the quadratic when $y = 0$.
- Because quadratic curves are symmetrical, the x coefficient of the turning point is halfway between the two roots.

Key Point

All quadratic graphs have a line of symmetry, which passes through the turning point.

The roots of a quadratic equation are the points where the graph crosses the x-axis. Not all quadratic curves will have roots.

Sketch the graph of equation $y = x^2 + 5x + 4$

Roots:

$x^2 + 5x + 4 = 0$

$(x + 4)(x + 1) = 0$

$x = -4$ or $x = -1$

y-intercept:

$y = 4$

y-intercept is (0, 4)

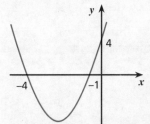

(−2.5, −2.25)

Turning Point:

$x = -4$ and $x = -1$, therefore $x = -2.5$

$y = (-2.5)^2 + (5 \times -2.5) + 4$

$\quad = -2.25$

minimum point is (−2.5, −2.25)

Substitute $x = 0$ into the equation.

Work out the values for x when $y = 0$.

The value of x for the turning point is in the middle of the two roots.

Substitute the value of x into the equation.

Transformation of the Graph $y = f(x)$

- Functions can be used to describe **transformations**.

Function	Transformation	
$f(x) + a$	Translation $\begin{pmatrix} 0 \\ a \end{pmatrix}$	Move up by a
$f(x) - a$	Translation $\begin{pmatrix} 0 \\ -a \end{pmatrix}$	Move down by a
$f(x + a)$	Translation $\begin{pmatrix} -a \\ 0 \end{pmatrix}$	Move left by a
$f(x - a)$	Translation $\begin{pmatrix} a \\ 0 \end{pmatrix}$	Move right by a
$-f(x)$	Reflection in x-axis	
$f(-x)$	Reflection in y-axis	

Key Point

If the coefficient of the x^2 term is **positive**, the graph will have a **minimum** point. If the coefficient of the x^2 term is **negative**, the graph will have a **maximum** point. The maximum and minimum point can also be found by completing the square (see p.97).

The curve $y = f(x)$ has a minimum point (2, −3).

Write down the coordinates of the turning point on the curve with equation:

a) $y = f(x + 2)$ The graph moves left by 2, so the minimum point is (0, −3).

b) $y = f(x) - 3$ The graph moves down by 3, so the minimum point is (2, −6).

c) $y = -f(x)$ The graph is reflected in the x-axis, so the maximum point is (2, 3).

Key Words

quadratic equation
quadratic graph
roots
intercept
turning point
maximum point
minimum point
transformation

Quick Test

1. Sketch the graph of the equation $y = x^2 - 3x + 2$
2. Draw the graph with the equation $y = x^2 - 6$
3. A quadratic function $y = f(x)$ has a maximum at the point (5, 3). Write down the maximum of the curve with equation
   a) $y = f(x) - 1$ and b) $y = f(x - 3)$

Powers, Roots and Indices

You must be able to:

- Recognise and recall powers of 2, 3, 4 and 5
- Recognise and recall the square numbers up to 15×15
- Calculate with powers and roots, including fractional and negative indices
- Estimate powers and roots of positive numbers
- Carry out exact calculations with surds.

Roots and Powers (Indices)

- Powers or **indices** are a shorthand method of showing that a number is multiplied by itself a number of times, e.g.
  $5 \times 5 = 5^2$
  $5 \times 5 \times 5 = 5^3$
  $5 \times 5 \times 5 \times 5 = 5^4$, etc.
- A **root** is the inverse function of a power.
- You must learn all the square numbers up to 15×15, i.e.
  1, 4, 9, 16, 25, 36, 49, 64, 81, 100, 121, 144, 169, 196, 225.
- You must also learn the cubes of 1, 2, 3, 4, 5 and 10, i.e.
  1, 8, 27, 64, 125 and 1000.
- You must be able to recognise powers of 2, 3, 4 and 5, e.g.
  $16 = 2^4$ and $243 = 3^5$, and work out real roots.

Work out the value of 3^4.	Write down the value of $\sqrt[3]{8}$.
$3 \times 3 \times 3 \times 3 = 81$	$2^3 = 8$, so $\sqrt[3]{8} = 2$

> **Key Point**
>
> $x^0 = 1$

Estimating Powers and Roots

- Roots of positive integers are either integers, fractions or irrational.
- Roots that are not integers or fractions can be estimated.

Between which two positive integers does $\sqrt{40}$ lie?

$6^2 = 36$ and $7^2 = 49$, therefore $\sqrt{40}$ is between 6 and 7. ◄——

> **Key Point**
>
> A square root can be both positive or negative.

As a decimal $\sqrt{40} = 6.325$ (to 3 d.p.)

Multiplying and Dividing Powers

- When completing calculations involving powers, apply the following rules:

$$x^m \times x^n = x^{m+n}$$
$$x^m \div x^n = x^{m-n}$$
$$(x^m)^n = x^{mn}$$
$$\frac{1}{x^n} = x^{-n}$$

Write each of the following as a single power of 2.

a) $2^3 \times 2^4 = 2^{3+4}$
$= 2^7$

b) $2^9 \div 2^4 = 2^{9-4}$
$= 2^5$

c) $(2^3)^4 = 2^{3 \times 4}$
$= 2^{12}$

d) $\dfrac{1}{2^5}$
$= 2^{-5}$

Negative and Fractional Powers

- When working with negative and fractional powers, the same basic rules apply.
- Fractional powers are used to express roots:
  - $x^{\frac{1}{2}} = \sqrt{x}$
  - $x^{\frac{1}{3}} = \sqrt[3]{x}$
  - $x^{\frac{1}{n}} = \sqrt[n]{x}$
  - $x^{\frac{m}{n}} = \left(\sqrt[n]{x}\right)^m$

a) Write down the value of $49^{\frac{1}{2}}$
$49^{\frac{1}{2}} = \sqrt{49}$
$= 7$

b) Write down the value of $16^{\frac{3}{4}}$
$16^{\frac{3}{4}} = \left(\sqrt[4]{16}\right)^3$
$= 2^3$
$= 8$

Surds

- **Surds** are square roots that are not integers or fractions and are, therefore, irrational.
- Surds can be manipulated using the rules below:
  - $\sqrt{ab} = \sqrt{a} \times \sqrt{b}$
  - $\sqrt{\dfrac{a}{b}} = \dfrac{\sqrt{a}}{\sqrt{b}}$

Simplify $\sqrt{200}$.

$\sqrt{200} = \sqrt{100 \times 2}$
$= \sqrt{100} \times \sqrt{2}$ ← Always make sure one of the numbers is a square number.
$= 10\sqrt{2}$

- It is bad practice to leave a surd on the denominator of a fraction.
- To find an equivalent fraction, **rationalise** the denominator.

Rationalise $\dfrac{1}{\sqrt{3}}$

$\dfrac{1}{\sqrt{3}} \times \dfrac{\sqrt{3}}{\sqrt{3}} = \dfrac{\sqrt{3}}{3}$ ← Multiply the numerator and denominator by $\sqrt{3}$.

Quick Test

1. Simplify $2x^3 \times 3x^2$
2. Write down the value of $25^{\frac{3}{2}}$
3. Rationalise $\dfrac{2}{\sqrt{2}}$

Area and Volume 1

You must be able to:

- Recall and use the formulae for the circumference and area of a circle
- Recall and use the formula for the area of a trapezium
- Recall and use the formulae for the volume and surface area of a prism
- Recall and use the formulae for the volume and surface area of a cylinder.

Circles

LEARN

Circumference of a Circle $(C) = 2\pi r$ or $C = \pi d$

Area of a Circle $(A) = \pi r^2$

> **Key Point**
>
> The symbol π represents the number **pi**.
>
> π can be approximated to 3.14 or $\frac{22}{7}$

Work out the circumference and area of a circle with radius 9cm. Give your answers to 1 decimal place.

Circumference
$C = 2 \times \pi \times 9$
$\quad = 18 \times \pi$
$\quad = 56.5$cm (to 1 d.p.)

Area
$A = \pi \times 9^2$
$\quad = \pi \times 81$
$\quad = 254.5$cm^2 (to 1 d.p.)

Trapeziums

LEARN

The area of a **trapezium** is:

$$A = \tfrac{1}{2}(a + b)h$$

where a and b are the **parallel** sides and h is the **perpendicular** height

> **Key Point**
>
> Perpendicular means 'at right angles'.
>
> Parallel means 'in the same direction and always the same distance apart'.

- This formula can be proved:

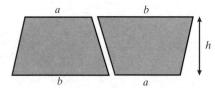

- Two identical trapeziums fit together to make a parallelogram with base $a + b$ and height h.
- The area of the parallelogram is $(a + b)h$.
- Therefore, the area of each trapezium is $\tfrac{1}{2}(a + b)h$.

> **Key Point**
>
> The area of a parallelogram is: $A = bh$
>
>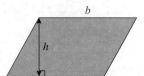

Work out the area of the trapezium.

$A = \tfrac{1}{2} \times (5 + 10) \times 4$

$\quad = 30$cm^2

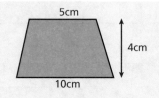

Prisms

- A right prism is a 3D shape that has the same **cross-section** running all the way through it.

LEARN

> Volume of a Prism = Area of Cross-Section × Length

- The surface area is the sum of the areas of all the **faces**.

Work out the volume and surface area of the triangular prism.

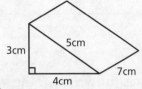

Volume
Area of the cross-section
$= \frac{1}{2} \times 3 \times 4 = 6\text{cm}^2$
Volume $= 6 \times 7$
$= 42\text{cm}^3$

Surface Area
Five faces:
Two triangular faces $= 6 + 6 = 12$
Base $= 4 \times 7 = 28$
Side $= 3 \times 7 = 21$
Slanted side $= 5 \times 7 = 35$
Total surface area:
$12 + 28 + 21 + 35 = 96\text{cm}^2$

Cylinders

LEARN

> Volume of a Cylinder $= \pi r^2 h$
>
> Surface Area of a Cylinder $= 2\pi rh + 2\pi r^2$

Work out the volume and surface area of the cylinder. Give your answers in terms of π.

Volume
$V = \pi \times 4^2 \times 7$
$= 112\pi\text{cm}^3$

Surface Area
$SA = 2 \times \pi \times 4 \times 7 + 2 \times \pi \times 4^2$
$= 56\pi + 32\pi$
$= 88\pi\text{cm}^2$

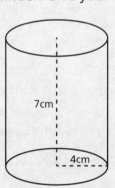

Key Point

A cylinder is just like any other right prism. To find the volume, you multiply the area of the cross-section (circular face) by the length of the cylinder.

Quick Test

1. Calculate the volume and surface area of a cylinder with radius 4cm and height 6cm.
2. Work out the area of the trapezium.

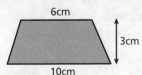

3. Calculate the circumference and area of a circle, diameter 7cm.

Key Words

trapezium
parallel
perpendicular
cross-section
face

Area and Volume 2

You must be able to:

- Find the volume of a pyramid
- Find the volume and surface area of a cone
- Find the volume of a frustum
- Find the volume and surface area of a sphere
- Find the area and volume of composite shapes.

Pyramids

- A **pyramid** is a 3D shape in which lines drawn from the **vertices** of the base meet at a point.

LEARN Volume of a Pyramid $= \frac{1}{3} \times$ Area of the Base $\times$ Height

Work out the volume of the square-based pyramid.

$V = \frac{1}{3} \times 9 \times 9 \times 7$

$= 189 \text{cm}^3$

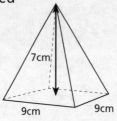

Cones

- A **cone** is a 3D shape with a circular base that tapers to a single vertex.

LEARN
Volume of a Cone $= \frac{1}{3}\pi r^2 h$

Surface Area of a Cone $= \pi r l + \pi r^2$

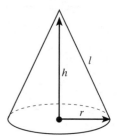

Work out **a)** the volume and **b)** the surface area of the cone.
Give your answers to 1 decimal place.

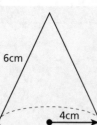

a) $h = \sqrt{6^2 - 4^2}$ ← First find the height using Pythagoras' Theorem.

$h = \sqrt{20}$

$V = \frac{1}{3} \times \pi \times 4^2 \times \sqrt{20} = 74.9 \text{cm}^3$

b) $SA = (\pi \times 4 \times 6) + (\pi \times 4^2) = 125.7 \text{cm}^2$

- A **frustum** is the 3D shape that remains when a cone is cut parallel to its base and the top cone removed.
- The original cone and the smaller cone that is removed are always similar.

LEARN
$$\frac{\text{Volume of}}{\text{a Frustum}} = \frac{\text{Volume of}}{\text{Whole Cone}} - \frac{\text{Volume of}}{\text{Top Cone}}$$

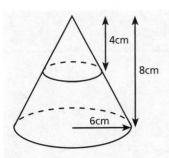

Calculate the volume of the frustum. Leave your answer in terms of π.

Radius of small cone = 3cm

$$V = \tfrac{1}{3}(\pi \times 6^2 \times 8) - \tfrac{1}{3}(\pi \times 3^2 \times 4)$$

$$= 84\pi \text{cm}^3$$

> The two cones are similar with scale factor 2.

Spheres

- A **sphere** is a 3D shape that is round, like a ball. At every point, its surface is equidistant from its centre.

> **LEARN**
>
> Volume of a Sphere = $\tfrac{4}{3}\pi r^3$
>
> Surface Area of a Sphere = $4\pi r^2$

- A **hemisphere** is half of a sphere; a dome with a circular base.

Work out **a)** the volume and **b)** the surface area of the sphere. Leave your answers in terms of π.

a) $V = \tfrac{4}{3} \times \pi \times 6^3 = 288\pi \text{cm}^3$

b) $SA = 4 \times \pi \times 6^2 = 144\pi \text{cm}^2$

Composite Shapes

Calculate the area of the shaded region.

$$A = (6 \times 7) - (\pi \times 1.5^2)$$

$$= 34.9 \text{cm}^2 \text{ (to 1 d.p.)}$$

> Find the area of the rectangle and subtract the area of the circle.

Work out the volume of the shape. Give your answer in terms of π.

Volume of the cylinder = $2.5^2 \times \pi \times 7.8 = \dfrac{195}{4}\pi \text{cm}^3$

Volume of the cone = $\tfrac{1}{3} \times \pi \times 2.5^2 \times 6.2 = \dfrac{155}{12}\pi \text{cm}^3$

Total volume = $\dfrac{195}{4}\pi + \dfrac{155}{12}\pi = \dfrac{185}{3}\pi \text{cm}^3$

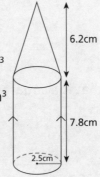

> **Key Point**
>
> To find the volume of a composite shape, you must break the shape down.

> **Quick Test**
>
> 1. Work out the volume of a sphere with diameter 10cm.
> 2. Calculate the surface area of a cone with radius 3cm and perpendicular height 6cm.
> 3. Work out the volume of a square-based pyramid with side length 5cm and perpendicular height 8cm.
> 4. Calculate the surface area of a hemisphere with radius 6cm.

> **Key Words**
>
> pyramid
> vertex / vertices
> cone
> frustum
> sphere
> hemisphere

Review Questions

Order and Value

1 An adult theatre ticket costs £38.60 and a child ticket costs £12.76.

Work out the total cost for 12 adult tickets and 5 child tickets. [2]

2 If $1.263 \times 2.47 = 3.11961$, work out:

   a) 12.63×0.247 [1]

   b) 0.1263×0.247 [1]

3 What is the value of $(-3)^5$? [1]

Total Marks _____ / 5

Types of Number

1 Use algebra to find three consecutive numbers that add up to 111. [3]

2 2, 9, 13, 15, 27, 35, 100
From the list of numbers above, find:

   a) A factor of 72. [1]

   b) A multiple of 7. [1]

   c) A square number. [1]

   d) A cube number. [1]

   e) A prime number. [1]

3 Find a number (other than 1) that is a square number and also a cube number. [1]

4 $25^2 \times 25^3 = 25^m$

Work out the value of m. [2]

5 Peter says $(2^3)^2 = (2^2)^3$

Is Peter right or wrong? Explain your answer. [2]

Total Marks _____ / 13

Basic Algebra & Factorisation and Formulae

1 There are k children in a room.

The number of children who wear glasses is g.

Write an expression in terms of k and g for the number of children who **do not** wear glasses. [1]

2 Simplify $7x - 2y + 5x - 3y$ [2]

3 Work out the value of the following expression when $a = -3$.

$$\frac{4a^2 - a^3}{a^4}$$ [1]

4 Expand and simplify $(y + 3)(2y - 7)(6 - y)$ [3]

5 Factorise $5ab - 3b^2c$ [1]

6 Explain why $5p - 7q$ cannot be factorised. [1]

7 Solve $\frac{x}{3} + 3 = 1$ [1]

8 The shape below is a rectangle.

$6x + 5$

$3x + 2$

Mitan thinks the correct expression for the perimeter of the rectangle is $9x + 7$.

a) Mitan is wrong. Explain his mistake. [1]

b) The perimeter of the rectangle is 56cm. Work out the value of x. [2]

9 The formula for the volume of a cylinder is $V = \pi r^2 h$.

a) Work out the volume of a cylinder with a radius (r) of 2cm and a height (h) of 10cm. [1]

b) Make r the subject of the formula. [2]

c) Work out the value of r when $V = 50$ and $h = 10$. [2]

Total Marks .... / 18

Ratio and Proportion

1 The square of the speed (v) at which a ball is thrown is directly proportional to the height (h) reached. A ball thrown at a speed of 10 metres per second reaches a height of 5 metres.

Calculate the height reached by a ball thrown at a speed of 30 metres per second. [4]

2 £60 is divided in the ratio 5 : 7

What is the difference in value between the two shares? [3]

3 Simplify 6.2 hours : 4 minutes [2]

Total Marks _____ / 9

Variation and Compound Measures

1 Martin is investing £400. He can choose between two ways of saving:

Type A – Simple interest at 6% per annum

Type B – Compound interest at 5% per annum

Which savings plan will give the better return after four years?
You **must** show your working. [4]

2 a) Calculate the distance travelled by a mouse moving at 1.5 metres per second for 1.5 seconds. [2]

b) Misty, a farm cat, can run at 11.25 miles per hour. Misty chases a mouse moving at 1.5 metres per second into a straight, plastic pipe.

If the mouse enters the pipe two seconds before the cat and the pipe is six metres long, would Misty catch the mouse before it escapes out of the other end? You **must** show your working. Assume 5 miles = 8 kilometres. [3]

3 It takes three dogs 15 days to eat a large sack of dog biscuits.

How long would it take five dogs to eat the same sized sack of dog biscuits at the same rate? [2]

Total Marks _____ / 11

Angles and Shapes 1 & 2

1 The three interior angles of a triangle are $y°$, $2y°$ and $3y°$.

Work out the size of the largest angle. [2]

2 A quadrilateral has one angle of 80°. Another angle is twice as big as the first angle and a third angle is 20° smaller than the first angle.

Work out the size of the fourth angle. [2]

3 This angle diagram is incorrect. Explain why. [1]

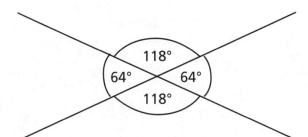

4 An aircraft flies from airport A on a bearing of 054° to airport B.

Work out the bearing that the aircraft must follow in order to return to airport A. [1]

5 Work out the exterior angle of a regular 15-sided shape. [2]

6 An irregular hexagon has interior angles in degrees of $2h$, $4h$, $4h$, $4h$, $5h$ and $5h$.

Work out the size of the smallest angle. [3]

7 State whether each of the following statements is **true** or **false**.

a) The sum of the interior angles of a heptagon is 900°. [1]

b) A parallelogram has no lines of symmetry and rotational symmetry of order 2. [1]

c) The direction south-east is on a bearing of 145°. [1]

Total Marks _____ / 14

Review Questions

Fractions

1 Work out $\frac{4}{25} + \frac{3}{5} \times \frac{2}{5}$ [2]

2 How many nails of length $\frac{8}{9}$ cm can be placed end to end on a 32cm line? [1]

3 What is half of a half of a quarter? [1]

4 A box contained 344 dog biscuits.
Fluff ate one-eighth of them.

How many biscuits were left? [2]

> **Total Marks** .... / 6

Percentages

1 A coat costs £56. It is reduced by 35% in a sale.

What is the sale price of the coat? [2]

2 Richard says that 30% of £40 is the same as 40% of £30.

Is he correct? Explain your answer. [2]

3 A house is purchased for £215 000 and sold for £300 000.

Calculate the percentage profit. [1]

4 On Black Friday, a coat was reduced by 35%.

If the sale price of the coat on Black Friday is £80, what was its original price to the nearest pound? [2]

> **Total Marks** .... / 7

Probability 1 & 2

1 The sides of a spinner are coloured red, blue, green and yellow.
The probability that the spinner will land on each colour is shown in the table below:

Colour	Red	Blue	Green	Yellow
Probability	0.3	0.3		0.1

   a) Complete the table to show the probability of landing on green. [1]

   b) Estimate the number of times the spinner will land on the colour red if it is spun 50 times. [2]

   c) Georgia and Annie are playing a game using the spinner.
     Georgia suggests that she wins if the spinner lands on red or blue and Annie wins if the
     spinner lands on green or yellow.

     Georgia thinks this is fair. Is she correct? Explain your answer. [2]

2 A restaurant serves three courses: starters, mains and desserts. All customers have a main
course. The Venn diagram shows information about customers who also had a starter, a dessert
or both.

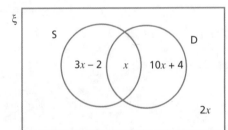

S = Starter
D = Dessert

The total number of customers served at lunchtime is 82.

Work out the number of customers who ordered neither a starter nor a dessert. [3]

3 Two events, A and B, are mutually exclusive: $P(A) = 0.3$ and $P(B) = 0.6$

   a) Draw a Venn diagram to represent this information. [2]

   b) Write down P(A and B) [1]

   c) Work out P(A or B) [2]

4 Helen has 10 yoghurts. Four are vanilla flavoured and six are banana flavoured.
She takes a yoghurt at random for breakfast on Monday and another on Tuesday.

Work out the probability that she takes one of each flavour. [4]

Total Marks _____ / 17

Practice Questions

Number Patterns and Sequences & Terms and Rules

1 Here is a sequence:

20, 16, 12, 8, 4 ...

Write down the expression for the nth term of the sequence. [2]

2 Work out the first five terms in the sequence with the nth term $3n - 5$. [2]

3 The first five terms of an arithmetic sequence are 14, 17, 20, 23, 26 ...

a) Write an expression for the nth term of this sequence. [2]

b) Calculate the 100th term in this sequence. [1]

4 The population of a culture of bacteria is given by the formula:
$P = 10 \times 2^t$, where t is the time in hours.

a) What is the initial population of bacteria? [2]

b) What is the size of the population after five hours? [2]

c) After how many hours will the population exceed 1000? [2]

d) Is this a sensible formula to model the population? Explain your answer. [1]

5 a) A sequence of numbers is given as 5, 8, 12, 19 ...

Write down the next two terms in the sequence. [2]

b) A second sequence of numbers is given as 2, 3, 2, 3, 2, 3 ...

Write down the 100th term. [1]

6 These are the first two patterns in a sequence made using matchsticks:

a) Draw the next two patterns in the sequence. [2]

b) Write down the rule for the number of matchsticks required for pattern number n. [2]

c) Use the rule to work out how many matchsticks are required for pattern 100. [1]

7 The first five terms of a quadratic sequence are:

4, 15, 32, 55, 84 ...

Write down the expression for the nth term of this sequence. [3]

Total Marks .... / 25

Transformations & Constructions

1 **a)** Plot the following points: $A(2, 0)$ $B(5, 0)$ $C(5, 2)$ $D(3, 2)$ $E(3, 5)$ $F(2, 5)$
Join the points together and label the shape M. [1]

b) Rotate shape M by 180° about the origin to form shape N. [1]

c) Reflect shape N in the *x*-axis to form shape O. [1]

d) Describe fully the single transformation that maps shape O to shape M. [2]

2 Rectangle R has a width of 3cm and a length of 5cm. It is enlarged by scale factor 3 to give rectangle T.

a) What is the area of rectangle T? [2]

b) How many times bigger is the area of rectangle T than the area of rectangle R? [2]

3 Describe how to construct an angle of 45°. [2]

4 Describe the locus of points in the following:

a) A person sitting on the London Eye as it rotates around. [1]

b) The seat of a moving swing. [1]

c) The end of the minute hand on a clock moving for one hour. [1]

d) The end of a moving see-saw. [1]

5 The diagram represents a solid made from 10 identical cubes.
On a squared grid, draw the:

a) Front elevation [1]

b) Plan view. [1]

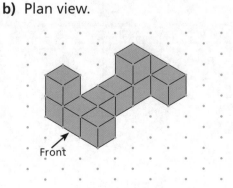
Front

Total Marks _____ / 17

Practice Questions

Linear Graphs & Graphs of Quadratic Functions

1 Draw the graph of $y = 4x - 2$ with values of x from –4 to 4.  [2]

2 Draw the graph of the function with gradient 5 and y-intercept (0, 3). [2]

3 Write down the gradient and y-intercept of the graph with equation $y = 5 - 2x$. [1]

4 Draw the graph with equation $y = 3x^2 - 2x + 1$ for values of x from –3 to 3. [2]

5 Work out the equation of the line that joins the points (–2, 5) and (3, –1). [3]

6 Work out the equation of the line drawn. [3]

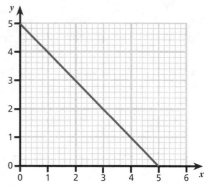

7 A straight line goes through the points (T, S) and (H, R).
$S = R - 5$ and $T = H + 2$

   a) What is the gradient of the line? [2]

   b) What is the y-intercept in terms of H and R? [1]

8 The graph has a minimum value at
the point (0, –8) and a maximum value at
the point (–2, –4).

Work out the new minimum and maximum
values after the following transformations.

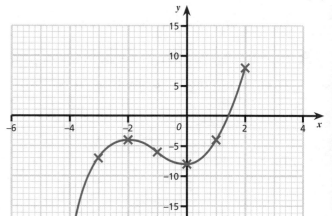

   a) $f(x + 2)$ [2]

   b) $f(-x)$ [2]

   c) $f(x) + 4$ [2]

   d) $f(x - 5)$ [2]

Total Marks .... / 24

Powers, Roots and Indices

1 Write down the following as a single power of 5:

a) $5^2 \times 5^3$ [1]

b) $5^7 \div 5^3$ [1]

c) $\dfrac{5^4 \times 5^2}{5^3}$ [2]

2 Write down the value of $16^{-\frac{1}{2}}$. [2]

3 Simplify $(x^2)^4$ [1]

4 Write the following in the form $a\sqrt{3} + b$ where a and b are integers.

$\left(2 + \sqrt{3}\right)\left(2 + 2\sqrt{3}\right)$ [3]

5 Simplify $\left(3r^2 p\right)^3$ [2]

> **Total Marks** _____ / 12

Area and Volume 1 & 2

1 Two identical circles sit inside a square of side length 6cm.

Work out the area of the shaded region. [4]

2 A vase is made from two cylinders. The larger cylinder has a radius of 15cm.
The total volume of the vase is 6000πcm³.
The ratio of volumes of the smaller cylinder to the larger cylinder is 1 : 3

a) Calculate the height of the larger cylinder. [3]

b) The height and radius of the smaller cylinder are equal. Work out the radius of the smaller cylinder. [3]

3 A cat's toy is made out of plastic. The top of the toy is a solid cone with radius 3cm and height 7cm. The bottom of the toy is a solid hemisphere. The base of the hemisphere and the base of the cone are the same size.

Calculate the volume of plastic needed to make the toy. Give your answer in terms of π. [3]

> **Total Marks** _____ / 13

Uses of Graphs

You must be able to:

- Use the form $y = mx + c$ to identify parallel and perpendicular lines
- Interpret the gradient of a straight-line graph as a rate of change
- Recognise and interpret graphs that illustrate direct and inverse proportion.

Parallel and Perpendicular Lines

- **Parallel** lines travel in the same direction and have the same gradient.
- **Perpendicular** lines are at right angles to each other.
- If a line has gradient m, then any line perpendicular to it will have a gradient of $-\frac{1}{m}$.

Write down the gradient of the line parallel to the line with the equation $y = 7 - 2x$.

The line has a gradient of –2 so the line that is parallel to it also has a gradient of –2.

> **Key Point**
>
> The gradient of a straight line in the form $y = mx + c$ is m.

Write down the gradient of the line that is perpendicular to the line with equation $y = 3x - 1$.

The line has a gradient of 3 so the line that is perpendicular to it has a gradient of $-\frac{1}{3}$.

Work out the equation of the line that goes through the point (2, 9) and is parallel to the line with equation $y = 7x + 10$.

$y = mx + c$

$y = 7x + c$ ← Substitute in $m = 7$

$9 = (7 \times 2) + c$ ← Goes through the point (2, 9), so $x = 2$ when $y = 9$

$c = -5$

The equation of the parallel line is $y = 7x - 5$.

Work out the equation of the line that is perpendicular to the line $y = \frac{3}{2}x + 2$ and goes through the point (10, 6).

$y = mx + c$

$y = -\frac{2}{3}x + c$ ← The line has a gradient of $\frac{3}{2}$ so the gradient of the perpendicular line is $-\frac{2}{3}$. This is your value for m.

$6 = (-\frac{2}{3} \times 10) + c$ ← Goes through the point (10, 6), so $x = 10$ when $y = 6$

$c = \frac{38}{3}$

The equation of the perpendicular line is
$y = -\frac{2}{3}x + \frac{38}{3}$ or $3y = 38 - 2x$

Gradient of a Line

- The **rate of change** is the rate at which one quantity changes in relation to another.
- The gradient of a straight-line graph represents a rate of change – it describes how the variable on the y-axis changes when the variable on the x-axis is increased by 1.

The graph below shows the volume of liquid in a container over time. What is the rate of change?

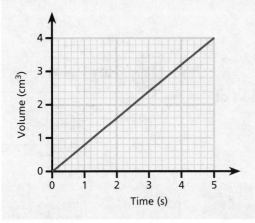

Gradient = $\frac{4}{5}$

$= \frac{4}{5} = 0.8\text{cm}^3/\text{s}$

Gradient = $\frac{\text{Change in } y}{\text{Change in } x}$

The gradient is the rate of change.

Real-Life Uses of Graphs

The graph below is the conversion graph between miles and kilometres.

a) How many kilometres are there in 5 miles?
   5 miles = 8km

   Read from the graph.

b) What is the gradient of the line?

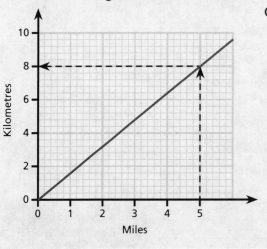

Gradient = $\frac{8}{5}$

$= 1.6$

1 mile = 1.6km

Quick Test

1. Work out the equation of the line that is parallel to the line $y = -2x + 6$ and goes through the point (4, 7).
2. Work out the equation of the line that is perpendicular to the line $y = -2x + 6$ and goes through the point (5, 1).

Key Words

parallel
perpendicular
rate of change

Other Graphs 1

You must be able to:

- Recognise, draw and interpret cubic, reciprocal and exponential graphs
- Interpret distance–time graphs and velocity–time graphs
- Work out acceleration from a velocity–time graph
- Work out speed from a distance–time graph.

Distance–Time Graphs

- A **distance–time graph** shows distance travelled in relation to a fixed point (starting point) over a period of time.
- The gradient of a straight line joining two points is the speed of travel between those two points.

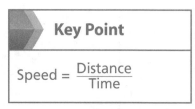

Key Point

$$Speed = \frac{Distance}{Time}$$

The graph below shows Val's car journey from St Bees to Cockermouth and back.

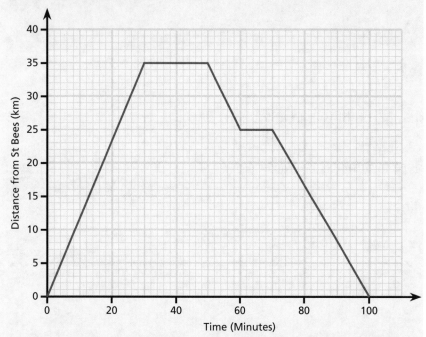

Distance from St Bees (km) / Time (Minutes)

a) Val sets off at 12 noon and travels directly to Cockermouth. At what time does she arrive?

Val travels for 30 minutes so arrives at 12.30pm.

b) For how long does Val stop in Cockermouth?

20 minutes ⟵ This is represented by the horizontal line on the graph – where the distance does not change.

c) Val begins her journey home but stops to fill up with petrol. Calculate the average speed of Val's journey from the petrol station to home in kilometres per hour.

$$Speed = \frac{Distance}{Time} = \frac{25}{0.5}$$ ⟵ Convert minutes into hours.

$$= 50km/h$$

Velocity–Time Graphs

- **Velocity** has both magnitude (size) and direction.
- The magnitude of velocity is called speed.
- The gradient of a straight line joining two points is the acceleration between those two points.
- Area Under Graph = Distance Travelled

The graph below shows part of the journey of a car.

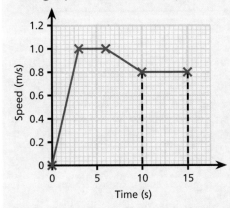

Time (s)

a) For how many seconds does the car decelerate?

4 seconds

b) What is the distance travelled in the last 5 seconds of the journey?

5 × 0.8 = 4m

c) What is the acceleration of the car in the first 3 seconds of the journey?

The acceleration is $\frac{1}{3}$ m/s²

Key Point

A positive gradient shows increasing speed.

A negative gradient shows decreasing speed.

Zero gradient shows constant speed.

Between 6 seconds and 10 seconds there is a negative gradient, so the car is decelerating.

This is the area of the rectangle under that part of the line.

This is the gradient of the line.

Cubic Function

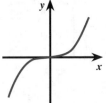

Reciprocal Function

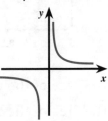

Exponential Function

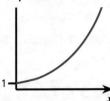

Other Graphs

- A **cubic function** is one that contains an x^3 term.
- A **reciprocal** function takes the form $y = \dfrac{a}{x}$
- An **exponential** function takes the form $y = k^x$

Quick Test

1. Plot the graph $y = 4^x$ for x values –4 to 4.
2. Plot the graph $y = 3x^3 - 5$ for values –2 to 2.
3. Below is a graph for the journey of a car.

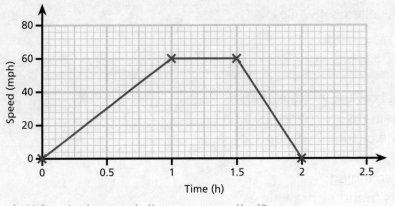

Time (h)

a) What is the total distance travelled?
b) For how many minutes is the car decelerating?

Key Words

distance–time graph
velocity
cubic function
reciprocal
exponential

Other Graphs 2

You must be able to:

- Estimate and interpret the area under a curve
- Work out and interpret a gradient at a point on a curve
- Work out the equation of a tangent to a circle.

Estimating the Area Under a Curve

- On a velocity–time graph, the area under the curve is equal to the distance travelled.
- To estimate the area under a curve, split the area into triangles and trapeziums. Remember, this is only an estimate.

Below is the velocity–time graph for a journey.

Estimate the distance travelled.

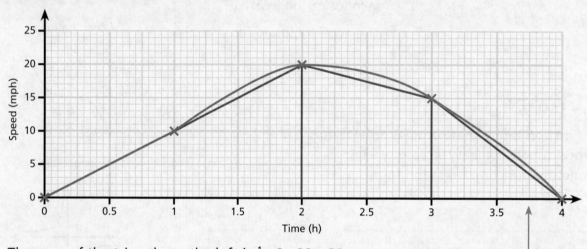

The area of the triangle on the left is $\frac{1}{2} \times 2 \times 20 = 20$

The area of the trapezium is $\frac{1}{2} \times (20 + 15) \times 1 = 17.5$

> Look carefully to check the units on the axes match, i.e. time in hours and speed in distance per hour.

The area of the triangle on the right is $\frac{1}{2} \times 1 \times 15 = 7.5$

The total area is 20 + 17.5 + 7.5 = 45

An estimate of the total distance travelled is 45 miles.

Rates of Change

- The **gradient** of a straight line describes how the variable on the y-axis changes when the variable on the x-axis is increased by 1 unit.
- On a distance–time graph, speed is the change in distance per unit of time. This is called a rate of change.
- The gradient of a curve is constantly changing.
- The gradient at any particular point is found by drawing a **tangent** at that point.

Work out the gradient of the curve $y = x^2$ at the point where $x = 3$.

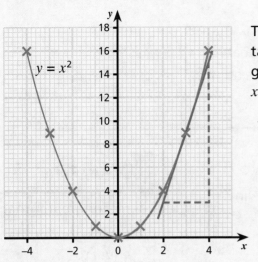

The gradient of the tangent is $\frac{12}{2} = 6$, so the gradient of the curve at $x = 3$ is 6.

Draw a tangent that meets the curve at the specified point.

Key Point

The gradient of the curve is the same as the gradient of the tangent at that point. A tangent is a straight line that touches the curve at a point.

Equation of a Circle

- The equation of a circle is of the form $x^2 + y^2 = r^2$ where the centre is (0, 0) and the radius is r.
- A circle with equation $x^2 + y^2 = 25$ has centre (0, 0) and radius 5.

Work out the equation of the tangent to the circle with equation $x^2 + y^2 = 25$ at the point (3, 4).

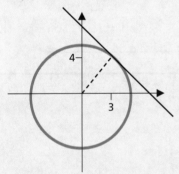

The gradient of the radius is $\frac{4}{3}$.

The gradient (m) of the perpendicular line, i.e. the tangent, is $-\frac{3}{4}$.

$4 = (-\frac{3}{4} \times 3) + c$

$c = \frac{25}{4}$

The equation of the tangent is $y = -\frac{3}{4}x + \frac{25}{4}$ OR $4y + 3x = 25$

Substitute $x = 3$, $y = 4$ and $m = -\frac{3}{4}$ into $y = mx + c$ to work out c.

Key Point

The radius and tangent of a circle meet at 90°.

Quick Test

1. Work out the equation of the tangent of the circle with equation $x^2 + y^2 = 29$ at the point (2, 5).
2. Plot the curve with equation $y = 3x^2 + 5$ and estimate the gradient at the point (2, 17).

Key Words

gradient
tangent

Inequalities

You must be able to:

- Solve linear inequalities in one or two variables
- Solve quadratic inequalities in one variable
- Represent solutions to inequalities on number lines or graphs.

Linear Inequalities

- The solution to an **inequality** can be shown on a number line.

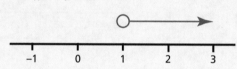

means $x <$ means $x >$

means $x \leqslant$ means $x \geqslant$

> **Key Point**
>
> $>$ means greater than
>
> $\geqslant$ means greater than or equal to
>
> $<$ means less than
>
> $\leqslant$ means less than or equal to

Solve these inequalities and show the solutions on a number line:

a) $x + 3 > 4$

$x > 4 - 3$

$x > 1$

b) $2(x + 4) \leqslant 18$

$x + 4 \leqslant 9$

$x \leqslant 5$

Work out all the possible integer values of n for these inequalities:

a) $-4 < n < 4$

$n = -3, -2, -1, 0, 1, 2, 3$

b) $-3 < 10n \leqslant 53$

$-0.3 < n \leqslant 5.3$ ← Divide each part of the inequality by 10.

$n = 0, 1, 2, 3, 4, 5$ ← n must be a whole number.

Graphical Inequalities

- The graph of the equation $y = 6$ is a line; the graph of the inequality $y > 6$ is a **region**, which has the line $y = 6$ as a boundary.
- For inequalities $>$ and $<$ the boundary line is **not included** in the solution and is shown as a **dashed line**.
- For inequalities $\geqslant$ and $\leqslant$ the boundary line is **included** in the solution and is shown as a **solid line**.

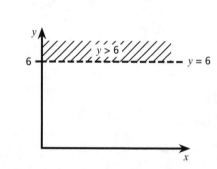

On a graph, show the region that satisfies $x \geqslant 0$, $x + y < 3$ and $y > 3x - 1$.

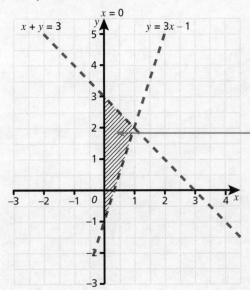

This is the region that satisfies all three inequalities.

Quadratic Inequalities

- A quadratic inequality contains an x^2 term and has a maximum of two possible solutions for x.
- To solve a quadratic inequality you must factorise the side with the quadratic.
- To determine the direction of the arrow for each solution, substitute values for x in the pairs of brackets.

Solve $x^2 + 2x - 15 \geqslant 0$ and show the solution on a number line.

$(x + 5)(x - 3) \geqslant 0$ ◄————

Factorise the expression.

If $(x + 5)(x - 3) = 0$, then $x = -5$ or 3

$(x + 5)(x - 3) \geqslant 0$ ◄————

Substitute values for x to determine the direction of the arrow.

If $x = -10$ then $(-10 + 5)(-10 - 3) = 65$, which satisfies $\geqslant 0$, so the arrow goes to the left from -5.

If $x = 10$, then $(10 + 5)(10 - 3) = 105$, which satisfies $\geqslant 0$, so the arrow goes to the right from 3.

So the solutions are $x \geqslant 3$ and $x \leqslant -5$.

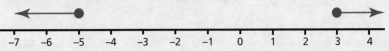

Quick Test

1. Solve the inequality $2x - 5 < 9$ and show the solution on a number line.
2. Solve the inequality $5x + 4 < 3x + 10$
3. Solve $x^2 + 5x - 14 \geqslant 0$

Key Words

inequality
region

Congruence and Geometrical Problems

You must be able to:

- Identify congruent and similar shapes
- State the criteria that congruent triangles satisfy
- Solve problems involving similar figures
- Understand geometrical problems.

Congruent Triangles

- If two shapes are the same size and shape, they are **congruent**.
- Two triangles are congruent if they satisfy one of the following four criteria:
  - SSS – three sides are the same
  - SAS – two sides and the **included angle** (the angle between the two sides) are the same
  - ASA – two angles and one corresponding side are the same
  - RHS – there is a right angle, and the hypotenuse and one other corresponding side are the same.
- Sometimes angles or lengths of sides have to be calculated before congruency can be proved.

State whether these two triangles are congruent and give a reason for your answer.

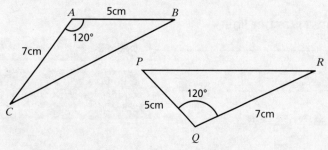

Angle CAB = Angle PQR (given)
$AC = QR$ (given)
$AB = PQ$ (given)
Triangles ABC and PQR are congruent because they satisfy the criteria SAS.

> **Key Point**
>
> Congruent shapes can be reflected, rotated or translated and remain congruent.

Similar Triangles

- **Similar** figures are identical in shape but can differ in size.
- In similar triangles:
  - corresponding angles are identical
  - lengths of corresponding sides are in the same ratio $y : z$
  - the area ratio = $y^2 : z^2$
  - the volume ratio = $y^3 : z^3$.

Triangles AED and ABC are similar.

Calculate **a)** AC and **b)** DC.

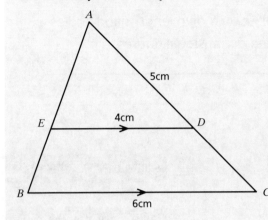

a) $\dfrac{4}{6} = \dfrac{5}{AC}$ ← The corresponding sides of both triangles are in the same ratio.

$4 \times AC = 6 \times 5$ ← Cross multiply.

$AC = \dfrac{6 \times 5}{4}$

$= 7.5\text{cm}$

b) $DC = AC - AD$

$= 7.5 - 5$

$= 2.5\text{cm}$

Geometrical Problems

- Congruency and similarity are used in many geometric proofs.

Prove that the base angles of an isosceles triangle are equal.

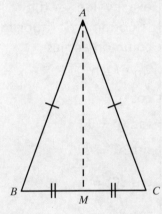

Given $\triangle ABC$ with $AB = AC$
Let M be the midpoint of BC
Join AM
$AB = AC$ (given)
$BM = MC$ (from construction)
$AM = AM$ (common side)
$\triangle ABM$ and $\triangle ACM$ are congruent (SSS)
So, angle ABC = angle ACB

Key Point

When writing a proof, always give a reason for each statement.

Quick Test

1.

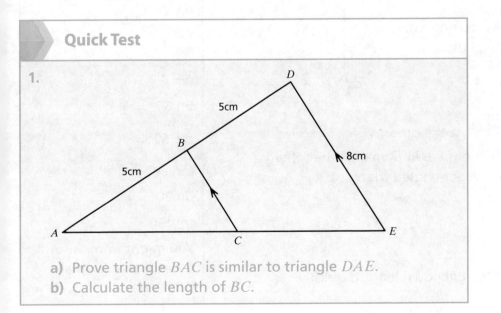

a) Prove triangle BAC is similar to triangle DAE.
b) Calculate the length of BC.

Key Words

congruent
included angle
similar

Right-Angled Triangles

You must be able to:

- Recall and use the formula for Pythagoras' Theorem
- Recall the trigonometric ratios
- Use the trigonometric ratios to calculate unknown lengths and angles in right-angled triangles in two and three-dimensional figures.

Pythagoras' Theorem

- **Pythagoras' Theorem** states $a^2 + b^2 = c^2$.
- The longest side (c) is opposite the right angle and called the **hypotenuse**.

LEARN

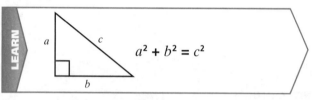

$$a^2 + b^2 = c^2$$

Careful: b is a shorter side so the answer must be **less** than 14.1cm.

Unless stated otherwise, give the answer to 3 significant figures.

Work out the length of the hypotenuse (c).

$3^2 + 4^2 = c^2$
$9 + 16 = c^2$
$c = \sqrt{25}$
$c = 5m$

Work out the length of b.

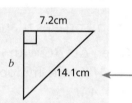

$14.1^2 - 7.2^2 = b^2$
$198.81 - 51.84 = b^2$
$146.97 = b^2$
$b = \sqrt{146.97}$
$b = 12.1cm$

A boat sails 15km due north, then 10km due east.

How far is the boat from its starting point? Give your answer to 3 decimal places.

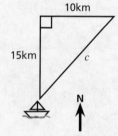

$15^2 + 10^2 = c^2$
$225 + 100 = c^2$
$c = 18.028km$

- The following length combinations are known as Pythagorean triples and regularly appear in right-angled triangles:
  - (3, 4, 5)
  - (6, 8, 10)
  - (5, 12, 13)
  - (7, 24, 25).
- Memorise them to help identify unknown lengths quickly.

Key Point

You need to know the exact values of the trigonometric ratios for common angles:

$\sin 0° = 0$

$\cos 0° = 1$

$\tan 0° = 0$

$\sin 30° = \frac{1}{2}$

$\cos 30° = \frac{\sqrt{3}}{2}$

$\tan 30° = \frac{1}{\sqrt{3}}$

$\sin 45° = \frac{1}{\sqrt{2}}$

$\cos 45° = \frac{1}{\sqrt{2}}$

$\tan 45° = 1$

$\sin 60° = \frac{\sqrt{3}}{2}$

$\cos 60° = \frac{1}{2}$

$\tan 60° = \sqrt{3}$

$\sin 90° = 1$

$\cos 90° = 0$

You might find it useful to learn the decimal values too.

ABCDEFGH is a cuboid.
Calculate the length of *AH* to 2 decimal places.

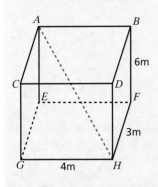

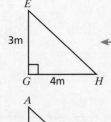

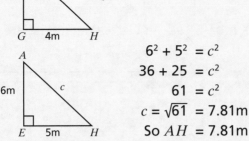

$$6^2 + 5^2 = c^2$$
$$36 + 25 = c^2$$
$$61 = c^2$$
$$c = \sqrt{61} = 7.81\text{m}$$
So *AH* = 7.81m

This is a Pythagorean triple, so
EH = 5m.

Key Point

a^2 means $a \times a$ **not** $2 \times a$.

Key Point

Sine, cosine and tangent ratios can **only** be used in right-angled triangles.

Trigonometric Ratios

* You can calculate unknown sides or angles in right-angled triangles using **sine**, **cosine** and **tangent**.
* θ is a Greek letter called **theta**. It stands for the unknown angle.

$$\sin \theta = \frac{\text{Opposite}}{\text{Hypotenuse}} \quad \cos \theta = \frac{\text{Adjacent}}{\text{Hypotenuse}} \quad \tan \theta = \frac{\text{Opposite}}{\text{Adjacent}}$$

* The above formulae can be remembered by:

Some **O**ld **H**orses **C**arry **A** **H**eavy **T**on **O**f **A**pples

SOH   **CAH**   **TOA**

* For example, $\sin \theta$ (**S**ome) $= \dfrac{\text{Opposite (\textbf{O}ld)}}{\text{Hypotenuse (\textbf{H}orses)}}$

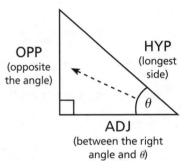

OPP
(opposite the angle)

HYP
(longest side)

θ

ADJ
(between the right angle and θ)

Work out x to 1 decimal place.

In this example you are looking for a missing length.

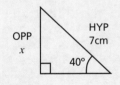

$$\sin 40° = \frac{x}{7}$$
$$x = 7 \times \sin 40°$$
$$x = 7 \times 0.6428 = 4.499...$$
$$x = 4.5\text{cm}$$

In this example you are looking for a missing angle.

Work out θ to the nearest degree.

On the calculator press either
[SHIFT] [tan] [1] [.] [6] or
[2ndF] [tan] [1] [.] [6]

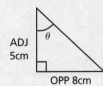

$$\tan \theta = \frac{8}{5} = 1.6$$
$$\tan^{-1} 1.6 = 57.99°$$
$$\theta = 58°$$

Key Words

Pythagoras' Theorem
hypotenuse
sine
cosine
tangent
theta
opposite
adjacent

> **Quick Test**
>
> 1. An isosceles triangle has side lengths of 7cm, 7cm and 5cm. Calculate the angle between the two equal sides to the nearest degree.
> 2. A triangle has sides 8cm, 15cm and 17cm. Is it right-angled?
> 3. *ABC* is a right-angled triangle, where angle *B* = 90°.
>   *AB* = 6cm, *AC* = 9cm. Calculate *BC* to 2 decimal places.

Sine and Cosine Rules

You must be able to:

- Recall and use the sine rule to work out an unknown side or angle of a triangle
- Recall and use the cosine rule to work out an unknown side or angle of a triangle
- Recall and use Area $= \frac{1}{2}ab \sin C$ to calculate the area, sides or angles of a triangle.

Solving Any Triangle

- The **sine rule** is used to calculate unknown angles or side lengths in triangles that are not right-angled.

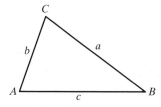

Key Point

Capital letters are used to represent angles; lower case letters are used to represent sides.

LEARN

Sine Rule: $\dfrac{a}{\sin A} = \dfrac{b}{\sin B} = \dfrac{c}{\sin C}$

Calculate the size of angle C.

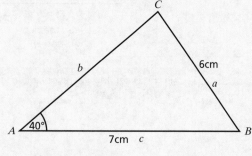

$$\frac{6}{\sin 40°} = \frac{7}{\sin C}$$

$6 \times \sin C = 7 \times \sin 40°$ ← Rearrange.

$\sin C = \dfrac{(7 \times \sin 40°)}{6}$

$\sin C = \dfrac{4.499513268}{6}$

$\sin C = 0.749918878$

$C = \sin^{-1} 0.749918878$ ← Always check your calculator is in degree mode.

$C = 48.58°$

Calculate the length of side BC to 2 decimal places.

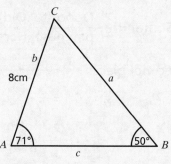

$$\frac{a}{\sin 71°} = \frac{8}{\sin 50°}$$ ← BC is side a.

$a \times \sin 50° = 8 \times \sin 71°$

$a = \dfrac{(8 \times \sin 71°)}{\sin 50°}$

$a = \dfrac{(8 \times 0.9455185756)}{0.7660444431}$

$a = \dfrac{7.564148605}{0.7660444431}$

$a = 9.87\text{cm (to 2 d.p.)}$

- The **cosine rule** is used for triangles without a right angle when:
    - two sides and the included angle (between them) are known
    - three sides are known.

LEARN

Cosine Rule: $a^2 = b^2 + c^2 - 2bc \cos A$ **OR** $\cos A = \dfrac{\left(b^2 + c^2 - a^2\right)}{2bc}$

Calculate the size of angle A to 1 decimal place.

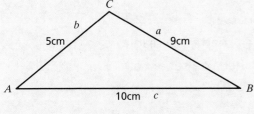

$$\cos A = \frac{(b^2 + c^2 - a^2)}{2bc}$$

$$\cos A = \frac{(5^2 + 10^2 - 9^2)}{2 \times 5 \times 10}$$

$$\cos A = \frac{44}{100}$$

$$\cos A = 0.44$$

Angle $A = \cos^{-1} 0.44$

Angle $A = 63.9°$

Calculate the length of side BC to 2 decimal places.

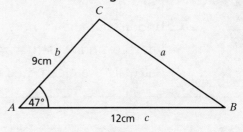

$a^2 = b^2 + c^2 - 2bc \cos A$ ← BC is side a.

$a^2 = 9^2 + 12^2 - 2 \times 9 \times 12 \times \cos 47°$

$a^2 = 225 - 216 \times \cos 47°$

$a^2 = 225 - 147.31$

$a^2 = 77.69$

$a = \sqrt{77.69}$

$a = 8.81$cm (to 2 d.p.)

Using Sine to Calculate the Area of a Triangle

- This formula for calculating the area of any triangle can be rearranged to work out unknown sides and angles:

Area $= \frac{1}{2}ab \sin C = \frac{1}{2}bc \sin A = \frac{1}{2}ac \sin B$

Key Point

If Area $= \frac{1}{2}ab \sin C$, then

$$a = \frac{2 \times \text{Area}}{b \sin C} \text{ and}$$

$$b = \frac{2 \times \text{Area}}{a \sin C}$$

$$\sin C = \frac{2 \times \text{Area}}{ab}$$

The area of the triangle ABC is 35.27cm². Calculate the length of AC. Give your answer to the nearest whole number.

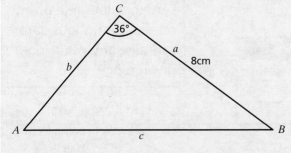

Area $= \frac{1}{2}ab \sin C$ ← Rearrange to make b (AC) the subject.

$$\frac{2 \times \text{Area}}{a \sin C} = b$$

$$\frac{2 \times 35.27}{8 \times \sin 36°} = b$$

$$\frac{70.54}{4.7023} = b$$

$$b = AC = 15\text{cm}$$

Quick Test

1. In triangle ABC, angle $A = 44°$, angle $C = 64°$ and $BC = 4$cm. Calculate a) the length of AB and b) the area of triangle ABC to 3 decimal places.

Key Words

sine rule
cosine rule

Statistics 1

You must be able to:

- Understand how a sample can be used to represent a population and its limitations
- Understand and identify different types of data
- Interpret and construct appropriate tables, charts and diagrams for data
- Calculate the mean, median, mode and range of a set of data.

Sampling

- A **population** is a collection of individuals or items.
- For some populations, collecting information from all members is not practical, so information is obtained from a **sample**.
- There are advantages and limitations associated with sampling:

Advantages	Limitations
Quicker and cheaper than investigating whole population	Bias can occur
Can be impractical to investigate whole population	Different samples could give different results

- There are two main types of sample: **random** and **stratified**.
- In a stratified sample, the population is divided into subgroups. The sample includes members from all subgroups in the same proportions as they occur in the population.
- **Primary data** is collected by yourself or on your behalf.
- **Secondary data** is collected from a different source.

Statistical Representation

- Data can be categorised as **qualitative** (non-numerical) or **quantitative** (numerical).
- Numerical data can be **discrete** or **continuous**.
- The type of diagram you use will be influenced by the data itself and what aspect of the data you want to look at.

Chart	Purpose
Pie chart	To show proportion but not exact frequencies
Bar chart	To compare frequencies
Line graph	To show changes in trends over time

Statistical Measures

- To compare data sets you should compare:
  - a measure of average: **mean**, **median** or **mode**
  - a measure of spread, i.e. the **range** – the difference between the largest and smallest value in the data set.
- A stem and leaf diagram orders data from smallest to largest and can be used to find the median and range of discrete data.

Key Point

Limitations can be minimised by ensuring the sample is large enough and representative of the whole population.

Key Point

In a stratified sample, the sample size for each subgroup is calculated using the formula:

$$\frac{\text{Size of Subgroup}}{\text{Size of Population}} \times \text{Sample Size}$$

Key Point

Discrete data takes certain values in a given range. Continuous data can take any value in a given range.

The Ages of Delegates at a Conference

2	1	1	2	3	4	4
3	2	2	2	3	5	8
4	3	3	5	8		
5	1	5	6			

Key: 2 | 1 = 21 years

The table shows the times taken by students, in minutes, to complete a mathematical puzzle.

Time Taken to Complete Puzzle (mins)	Frequency (f)	Cumulative Frequency	Midpoint (x)	fx
$0 < t \leqslant 2$	8	8	1	8
$2 < t \leqslant 4$	12	20	3	36
$4 < t \leqslant 6$	10	30	5	50
$6 < t \leqslant 8$	5	35	7	35
	$\Sigma f = 35$			$\Sigma fx = 129$

a) Write down the modal class.

The modal class is $2 < t \leqslant 4$.

b) Work out which group contains the median.

$\frac{35 + 1}{2}$ = 18th value

The median (18th value) is in the group $2 < t \leqslant 4$.

c) Estimate the mean time taken.

Mean = $\frac{129}{35}$ = 3.69 minutes (to 2 d.p.)

> **Key Point**
>
> Σ means 'sum of'.

The modal class has the highest frequency.

The median is the middle value.

There are 35 students.

Mean = $\frac{\Sigma fx}{\Sigma f}$

Scatter Diagrams

- **Scatter diagrams** are used to investigate the relationship between two variables.
- If a linear relationship exists, a line of best fit can be drawn. This can be used to make predictions.
- A prediction taken from the line of best fit within the data range is reliable. Taken from outside the data range, it may not be reliable.
- **Positive correlation** – as one variable increases, the other variable increases.
- **Negative correlation** – as one variable increases, the other variable decreases.
- You can describe correlation as weak or strong.

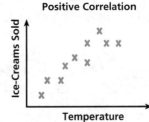

Positive Correlation

As temperature increases, ice-cream sales increase

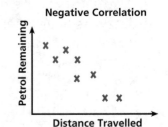

Negative Correlation

As distance travelled increases, petrol remaining decreases

> **Key Words**
>
> population
> sample
> random
> stratified
> primary data
> secondary data
> qualitative
> quantitative
> discrete data
> continuous data
> mean
> median
> mode
> range
> scatter diagram
> positive correlation
> negative correlation

> **Quick Test**
>
> 1. The table below shows the age of children who attend a reading group at the library. Calculate a) the mean b) the median c) the mode and d) the range of the data.
>
> | Age | 6 | 7 | 8 | 9 |
> |---|---|---|---|---|
> | Frequency | 5 | 10 | 6 | 12 |
>
> 2. Write down one advantage and one disadvantage of taking a sample instead of surveying the whole population.

Statistics 2

You must be able to:

- Construct and interpret diagrams for grouped data, including frequency polygons, histograms and cumulative frequency curves
- Construct and interpret box plots
- Compare data sets using quartiles and interquartile range.

Frequency Polygons

- A **frequency polygon** can be used to represent **grouped data**.
- A point is plotted for the midpoint of each group and joined by straight lines.

Cumulative Frequency Graphs

- The **cumulative frequency** is the running total. It is found by adding all the frequencies together.
- A cumulative frequency curve can be used to estimate the **median** and **interquartile range**.

Draw a cumulative frequency curve to represent the following data. Use your graph to estimate the median and interquartile range.

Time Taken to Run 100m (s)	Frequency	Cumulative Frequency
$12 < t \leqslant 14$	4	4
$14 < t \leqslant 17$	21	(4 + 21 =) 25
$17 < t \leqslant 19$	14	(25 + 14 =) 39
$19 < t \leqslant 23$	9	(39 + 9 =) 48

Work out the cumulative frequency and plot against the value at the end of the group.

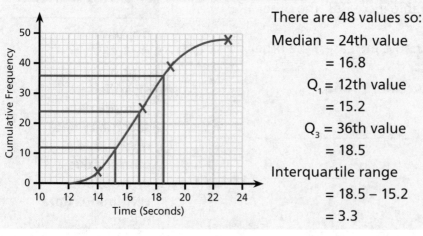

There are 48 values so:

Median = 24th value

= 16.8

Q_1 = 12th value

= 15.2

Q_3 = 36th value

= 18.5

Interquartile range

= 18.5 – 15.2

= 3.3

Histograms and Box Plots

- A **histogram** is used to represent **continuous data**.
- The area of each bar on a histogram is proportional to the frequency for that class.
- The height of the bar is equal to the frequency density.

Construct a histogram to represent the data
in the previous example for time taken to run 100m.

Time Taken to Run 100m (s)	Frequency	Class Width	Frequency Density
$12 < t \leqslant 14$	4	2	(4 ÷ 2 =) 2
$14 < t \leqslant 17$	21	3	(21 ÷ 3 =) 7
$17 < t \leqslant 19$	14	2	(14 ÷ 2 =) 7
$19 < t \leqslant 23$	9	4	(9 ÷ 4 =) 2.25

> **Key Point**
>
> Frequency Density
> $= \dfrac{\text{Frequency}}{\text{Class Width}}$

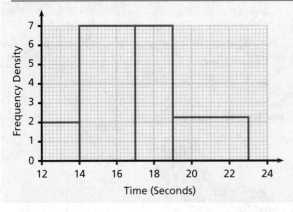

- A **box plot** displays information about the range, median and quartiles of a data set. It must be drawn against a scale.

Draw a box plot for the data for the time taken to run 100m.

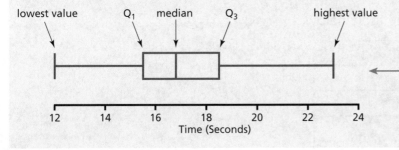

Use the values taken from the cumulative frequency graph.

Quick Test

1. For the following data construct a histogram.

Length of Leaf (cm)	Frequency
$5 < l \leqslant 7$	9
$7 < l \leqslant 12$	17
$12 < l \leqslant 20$	23

2. The box plot below shows information about the length of worms.
   a) Write down the median.
   b) Work out an approximate value for the interquartile range.

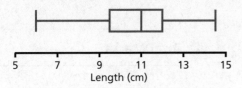

> **Key Words**
>
> **frequency polygon**
> **cumulative frequency**
> **median**
> **interquartile range**
> **quartile**
> **histogram**
> **box plot**

Number Patterns and Sequences & Terms and Rules

1 The first term that the following two sequences have in common is 17.

8, 11, 14, 17, 20 …

1, 5, 9, 13, 17 …

Work out the next term that the two sequences have in common.
You must show your working. [2]

2 Regular pentagons of side length 1cm are joined together to make a pattern.

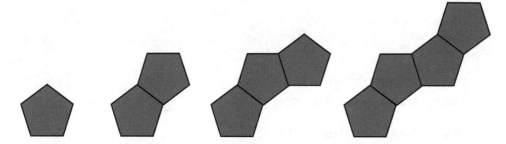

a) Use the patterns to complete the table below.

Pattern Number	Perimeter (cm)
1	
2	
3	
4	
60	
n	

[2]

b) What is the maximum number of pentagons that could be used to give a perimeter less than 1500cm? [2]

3 The nth term of a sequence is $n^2 + 6$.
Jenny states that 'the nth term is odd if n is odd and the nth term is even if n is even'.

Is Jenny correct? Explain your answer. [2]

4 Write down the nth term for the sequence below:

4, 6, 10, 18, 34 … [1]

Total Marks _____ / 9

Transformations & Constructions

1 Three points $X(5,1)$, $Y(3, 5)$ and $Z(1, 2)$ are reflected in the y-axis.

   a) Give the new coordinates of the three points. [3]

   b) The original points X, Y and Z are rotated 90° about (0, 0) in a clockwise direction.

      Give the coordinates of the three points in their new positions. [3]

2 A rectangle (C) measures 3cm by 5.5cm. Each length of rectangle C is enlarged by scale factor 3 to form a new rectangle (D).

   What is the ratio of the area of rectangle C to rectangle D? [3]

3 Describe the locus of points for the following:

   a) The path of a rocket for the first three seconds after take-off. [1]

   b) A point just below the handle on an opening door. [1]

   c) The central point of a bicycle wheel as the bicycle travels along a level road. [1]

   d) The end of a pendulum on a grandfather clock. [1]

4 Describe the plan view of a cube measuring 4cm by 4cm by 4cm. [1]

5 A can of baked beans has a circular lid of circumference 22cm and a height of 8cm.

   Draw the side elevation of the tin when it is standing upright. (Take π as $\frac{22}{7}$) [2]

6 The photograph shows a World War II Lancaster Bomber.

   Sketch:

   a) The side elevation of the Lancaster Bomber [2]

   b) The front elevation of the Lancaster Bomber [2]

   c) The plan view of the Lancaster Bomber. [2]

Total Marks _____ / 22

Linear Graphs & Graphs of Quadratic Functions

1 Work out the equation of the line that joins the points $\left(\frac{2}{3}, 8\right)$ and $\left(\frac{5}{6}, 5\right)$. [3]

2 Work out the equation of the line drawn below. [3]

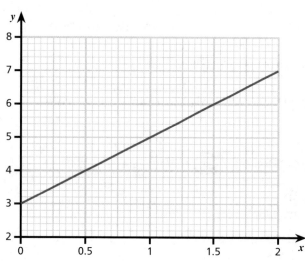

3 Sketch the graph of the function $y = x^2 + 4x + 3$, clearly stating the roots and the coordinates of the turning point. [3]

4 The equation of a line is $4y = 3x + 1$

Work out the gradient and y-intercept of the line. [2]

5 A curve has the equation $y = x^2 + ax + b$
The curve crosses the x-axis at the points (–7, 0) and (1, 0).

a) Work out the values of a and b. [3]

b) Work out the coordinates of the turning point. [2]

6 **a)** Sketch the graph $y = \frac{1}{x}$ [1]

b) On the same axes sketch the graph $y = -\frac{1}{x}$ [1]

c) Describe two different transformations that will transform the graph in part **a)** to the graph in part **b)**. [2]

Total Marks _____ / 20

Powers, Roots and Indices

1 Simplify $\left(2ab^{-5}\right)^{-3} \times \left(3a^{-2}b^3\right)^2$ [3]

2 Write $\frac{2}{\sqrt{3}} - \frac{6}{\sqrt{27}} + 6\sqrt{48}$ in the form $a\sqrt{3}$, where a is an integer. [3]

3 Amber states that $(x^{-2})^3 = \frac{1}{x^6}$

Is Amber's statement true or false? Explain your answer. [2]

4 Work out the value of $25^{-\frac{1}{2}}$. [1]

5 $a < \sqrt{170} < b$, where a and b are integers. $b - a = 1$.

Work out the values of a and b. [2]

> **Total Marks** .... / 11

Area and Volume 1 & 2

1 **a)** Work out the volume of the triangular prism. [2]

b) A cube has the same volume as the triangular prism.

Work out the total length of all the edges of the cube. [3]

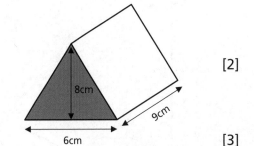

2 The numerical values of the area and circumference of a circle are equal.

Work out the radius of this circle. [2]

3 The volume of the trapezoid is 900cm³.
All measurements are in centimetres.

Work out the value of x. [4]

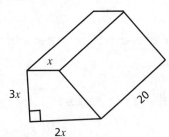

4 The surface area of a sphere is 75cm². Work out the length of the radius. [3]

> **Total Marks** .... / 14

Uses of Graphs & Other Graphs 1 & 2

1 A line is parallel to the line of equation $y = 3x - 2$ and goes through the point (1, 5).

Work out the equation of the line. [3]

2 Complete the table below.

Gradient of Line	Gradient of Parallel Line	Gradient of Perpendicular Line
5		
	3	
		$-\frac{1}{4}$
$\frac{8}{9}$		

[3]

3 Gemma, Naval and Esmai entered a five-mile cycling race.
The graph below shows the race.

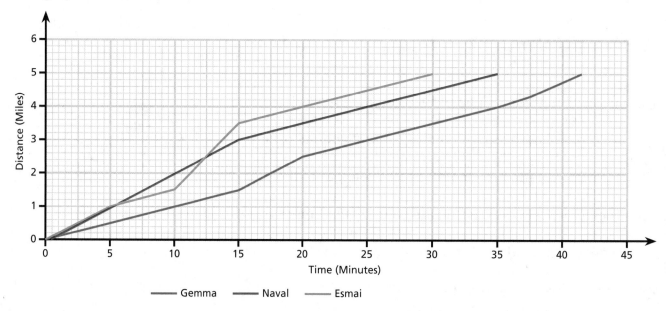

a) Who won the race? [1]

b) What speed was Naval travelling at for the last 20 minutes before he finished?
Give your answer in miles per hour. [2]

c) Between what times was Gemma travelling her fastest?
Give a reason for your answer. [2]

d) How many minutes after the race started did the winner move into the lead? [1]

e) Describe the race. [3]

Uses of Graphs & Other Graphs 1 & 2 (Cont.)

4 **a)** On the same set of axes, sketch the graphs of $y = 3^x$ and $y = 5^x$. [2]

b) Write down one property that the two graphs have in common. [1]

5 The graph below shows the conversion rate between British Pound Sterling (£) and the US Dollar ($).

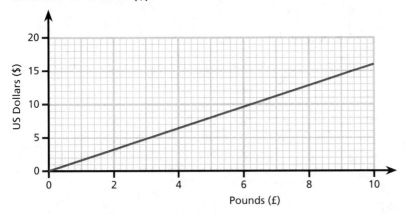

a) Calculate the gradient of the line. [2]

b) Interpret the gradient in terms of the exchange rate between British Pounds Sterling and US Dollars. [1]

6 The circle has equation $x^2 + y^2 = 41$

a) Work out the equation of the tangent to the circle at the point (4, –5). [4]

b) Write down the equation of a line parallel to the tangent. [1]

7 The graph below shows the journey of a car.

Work out the total distance travelled. [3]

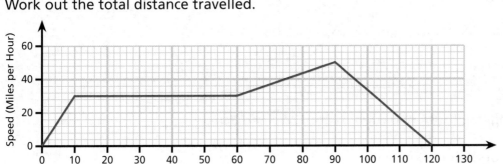

Total Marks _____ / 29

Inequalities

1. Solve $5 - 4x > 25$ and show the answer on a number line. [2]

2. Write down the inequality represented by the shaded region.

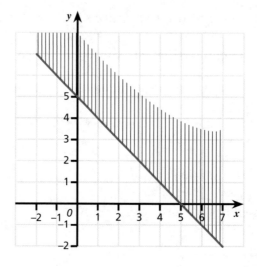

[1]

3. A bus can carry no more than 52 passengers.

Write down an inequality equation to show this information, where p is the number of passengers. [2]

Total Marks _____ / 5

Congruence and Geometrical Problems

1. Prove that triangle ABC and triangle BCD are similar. [3]

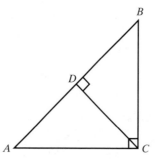

2. The corresponding lengths of two cuboids are 12cm and 3cm.

What is the ratio of their volumes? [1]

Total Marks _____ / 4

Right-Angled Triangles

1 A man walks 6.7km due north. He then turns due west and walks 7.6km.

How far is he now from his starting point? [3]

2 A 4m ladder leans against a vertical wire fence. The foot of the ladder is 2m from the base of the fence. Fang, the lion, can jump 3m vertically.

Will Fang be able to jump over the fence?
You must give reasons for your answer. [4]

3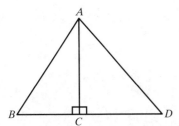
AB = 10cm, BC = 8cm and CD = 6cm.

a) Calculate AD to 3 significant figures. [3]

b) What type of triangle is triangle ABD? [1]

4 A bumblebee leaves its nest and flies 10 metres due south and then 6 metres due west.

What is the shortest distance the bumblebee has to fly to return to its nest?
Give your answer to 3 significant figures. [3]

5 A triangle has side lengths of 1.5cm, 2.5cm and 2cm. Is it a right-angled triangle?
Explain your answer. [3]

6 How long is the diagonal of a square of side length 3cm? [2]

7 A regular hexagon is inscribed in a circle of radius 6cm.

Use sine to prove that the side length of the hexagon is 6cm. [2]

8 A is the point (4, 0) and B is the point (7, 5).

Calculate the angle between AB and the x-axis to the nearest degree. [2]

9 A lifeguard is at the top of a lookout tower of height 14m, situated on a small island.
She sees a swimmer (P) due west of her at an angle of depression of 35°.
She sees another swimmer (Q) due south at an angle of depression of 18°.
Work out:

a) The distance of swimmer P from the base of the tower. [2]

b) The distance of swimmer Q from the base of the tower. [2]

c) The shortest distance between swimmers P and Q. Give your answer to 1 decimal place. [2]

Total Marks .... / 29

Sine and Cosine Rules

1 In a triangle ABC, AC = 5.7cm and AB = 7.5cm. The area of the triangle is 20cm².

Calculate angle CAB to the nearest degree. [4]

2 ABC is a triangle where BC = 12cm and CA = 14cm.

If angle ABC = 50°, calculate angle BAC to 2 decimal places. [3]

3 A tourist was standing 30m from the base of the Leaning Tower of Pisa.
The angle of elevation to the top of the tower was 30°.
The distance of the tourist from the top of the tower was 80m.

Calculate the slanting height (l) of the Leaning Tower of Pisa.
Give your answer to the nearest metre. [5]

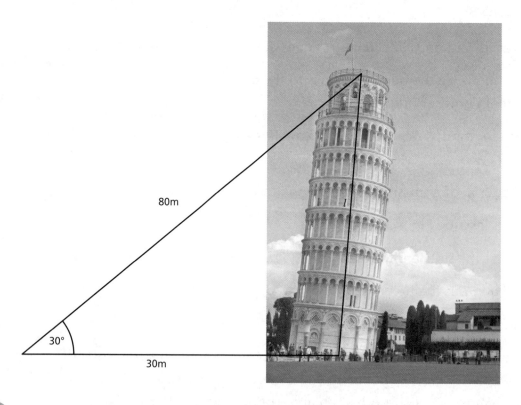

80m

30°

30m

4 In a golf championship, the distance from the tee to the third hole is 250m.
A golfer hits the ball 220m and it ends up 78m from the hole.

Calculate the angle (from the tee) that the ball went off course to 2 decimal places. [2]

Total Marks _____ / 14

Statistics 1 & 2

1 Below is a table that shows the times taken for a group of students to run 100m.

Time (seconds)	Frequency
$10 < t \leqslant 11$	6
$11 < t \leqslant 12$	21
$12 < t \leqslant 14$	13
$14 < t \leqslant 18$	10

a) How many students ran 100m? [1]

b) Construct a histogram for this data. [3]

c) Use your histogram to estimate the number of students who took between 11 and 15 seconds to complete the run. [2]

2 The cumulative frequency curve gives information about the test results for 150 students. The lowest mark was 10 and the highest was 75.

Use this information to draw a box plot to represent the data. [4]

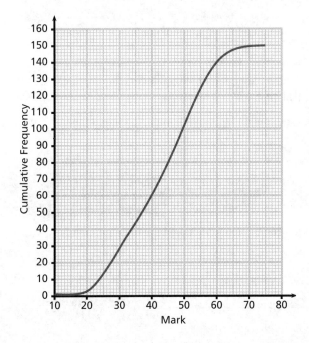

3 The table below shows hours of sunshine and amount of rainfall for nine towns across England in one month.

Sunshine (h)	600	420	520	630	470	380	560	430	450
Rainfall (mm)	11	18	13	9	16	25	14	20	19

a) Draw a scatter diagram to represent this information. [2]

b) Describe the relationship between hours of sunshine and rainfall. [1]

Total Marks _____ / 13

Measures, Accuracy and Finance

You must be able to:

- Understand and solve problems relating to household finance
- Check calculations using approximation
- Round numbers and measures to an appropriate degree of accuracy
- Apply and interpret limits of accuracy, including for questions set in context.

Solving Real-Life Problems

- Being able to understand and solve problems relating to finance is an essential life skill.

> Nadine bought a designer handbag from eBay® for £230 and sold it two years later for £184.
>
> Calculate the percentage loss.
>
> Percentage profit or loss = $\dfrac{\text{profit or loss}}{\text{original amount}} \times 100$
>
> $= \dfrac{£46}{£230} \times 100 = 20\%$

Loss = £230 – £184 = £46

> VAT is charged at the standard rate of 20%.
>
> What is the final cost after VAT has been added to a bill of £45?
>
> $\dfrac{20}{100} \times £45 = £9$
>
> Total bill = £45 + £9
>
> $= £54$

This could also be calculated using a multiplier:
1.2 × £45 = £54

> Yvonne earns £36 000 per year. The first £8200 is tax free, but income tax of 20% must be paid on the rest.
>
> How much income tax does Yvonne have to pay each year?
>
> £36 000 – £8200 = £27 800
>
> Income tax = $\dfrac{20}{100} \times £27\,800$
>
> $= £5560$

Work out the amount that is taxable.

Limits of Accuracy

- You will often be asked to give answers to a certain number of **decimal places** or **significant figures**. To do this, you must 'round off' the number.
- **Decimal places** refers to the number of digits after the decimal point, e.g.
  - 27.3652 = 27.4 (to 1 d.p.)
  - 27.3652 = 27.37 (to 2 d.p.)

Because the digit two places after the decimal point is 5 or more, the 3 (in the first decimal place) rounds up to 4.

The 5 rounds up the 6 to a 7.

- The first **significant figure** is the first non-zero figure (working from left to right), e.g.
  - 541 = 540 (to 2 significant figures)
  - 0.0347 = 0.035 (to 2 significant figures)

Approximation of Calculations

- You can use **approximations** to estimate and check the answers to calculations.

> Estimate the answer to (2136 + 39.7) ÷ (9.6 × 11.1)
>
> (2000 + 40) ÷ (10 × 10) ←
>
> = 2040 ÷ 100
>
> = 20.4

Approximate each number to 1 significant figure.

Problems Involving Limits of Accuracy

- Measurement can be approximate, e.g. a length (x) of 145cm to the nearest centimetre could be anything from 144.5cm to 145.499 99… cm
- Inequality notation is used to specify simple error intervals due to rounding, e.g. $144.5 \leqslant x < 145.5$
- When numbers are rounded off, the actual value lies between a **lower bound** and an **upper bound**. The bounds are called the **limits of accuracy**.

> A tennis court is measured and is found to be 23.4m by 10.2m. Each measurement is given to the nearest 10cm.
>
> Calculate the maximum and minimum possible values for the area of the tennis court.
>
> Maximum area = 23.45 × 10.25 ←
>
> = 240.3625m²
>
> Minimum area = 23.35 × 10.15 ←
>
> = 237.0025m²

Multiply the upper bounds.

Multiply the lower bounds.

Quick Test

1. Estimate the value of $(0.897)^2 \times 392.4$
2. A rectangle measures 2.81cm by 3.87cm to 2 decimal places. Write down the limits of accuracy in which each dimension lies.

Solving Quadratic Equations

You must be able to:

- Solve quadratic equations by factorising
- Solve quadratic equations by using a graph
- Solve quadratic equations by using the quadratic formula
- Solve quadratic equations by completing the square.

Factorisation

- When solving a quadratic equation by **factorisation** (see p.14–15), make sure it equals zero first.

Solve the equation $x^2 + 4x + 3 = 0$ by factorisation.

×	x	+1
x	x^2	$+x$
+3	$+3x$	$+3$

$(x + 1)(x + 3) = 0$

$x + 1 = 0 \qquad x + 3 = 0$

$\qquad x = -1 \qquad\quad x = -3$

> **Key Point**
>
> If two brackets have a product of zero, one of the brackets must equal 0.

Set up and complete a table. The missing terms need to have a product of +3 and a sum of +4.

First row = first bracket; first column = second bracket

The Method of Intersection

- Plotting a graph of a quadratic equation can give zero, one or two solutions for x when $y = 0$.
- The solutions are given by the curve's points of **intersection** with the x-axis. These solutions will sometimes be estimates.

> **Key Point**
>
> The points of intersection with the x-axis are called roots.

Find approximate solutions to the equation $x^2 - 5x + 1 = 0$ by plotting a graph.

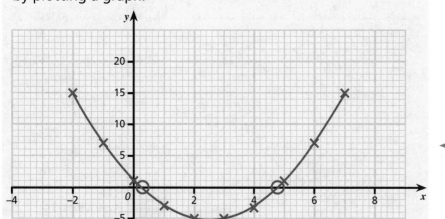

Draw the graph of $y = x^2 - 5x + 1$ (see p.48).

There are two solutions: $x = 0.2$ or $x = 4.8$

These solutions are approximate.

- This method can also be used to solve equations where the solutions are the points of intersection between a curve and a line.

Using the Quadratic Formula

- The quadratic formula is used to solve equations in the form
  $ax^2 + bx + c = 0$

The quadratic formula is $x = \dfrac{-b \pm \sqrt{b^2 - 4ac}}{2a}$

Using the quadratic formula, solve $4x^2 - x - 2 = 0$

$$x = \frac{1 \pm \sqrt{(-1)^2 - 4 \times 4 \times (-2)}}{2 \times 4}$$

$x = 0.843$ or -0.593

Key Point

Remember, when calculating b^2, a negative number squared is positive.

Substitute in: $a = 4$, $b = -1$, $c = -2$

Completing the Square

- To **complete the square**, you must write the equation in the form $(x + p)^2 + q = 0$ and then solve it.
- If the **coefficient** of the x^2 term is **not** 1, you must take the coefficient out as a factor and then complete the square.

Solve the equation $x^2 - 2x - 2 = 0$

$x^2 - 2x = 2$

$(x - 1)^2 - 1^2 = 2$

$(x - 1)^2 - 1 = 2$

$(x - 1)^2 = 3$

$x = 1 \pm \sqrt{3}$

Move the constant over to the right-hand side.

The number in the brackets is always half the coefficient of x.

Always minus the square of that number.

Solve.

Iteration

- **Iteration** is the repetition of a mathematical procedure.
- The results of one iteration are used as the starting point for the next.

Solve $x^2 - 3x - 1 = 0$ using iteration, to 2 decimal places.

$x = \dfrac{1}{x - 3}$ or $x = \dfrac{1}{x} + 3$

$x_{n+1} = \dfrac{1}{x_n - 3}$ or $x_{n+1} = \dfrac{1}{x_n} + 3$

The solutions are $x = -0.30$ and $x = 3.30$

Key Point

Iteration involves solving an equation many times. With each iteration your answer becomes more accurate.

Rearrange to make x the subject.

These two equations give the two solutions.

Substitute in any value for x_n. Use the results as a starting point for the next substitution. Continue until the terms converge.

Key Words

factorisation
intersection
complete the square
coefficient
iteration

Quick Test

1. Solve the equation $x^2 = 2x + 5$ by the method of intersection.
2. Solve the equation $3x^2 - 5x - 1 = 0$ using the quadratic formula.
3. Solve the equation $x^2 - 4x + 2 = 0$ by completing the square.
4. Solve $x^2 - 5x - 1 = 0$ using iteration (to 2 d.p.).

Simultaneous Equations and Functions

You must be able to:

- Solve linear and quadratic simultaneous equations algebraically
- Find approximate solutions to simultaneous equations using a graph
- Translate simple situations into two simultaneous equations and solve
- Work out composite functions.

Algebraic Methods

- **Linear simultaneous equations** can be solved by elimination.

Solve the following simultaneous equations:

$3x - y = 18$ Equation 1

$x + y = 10$ Equation 2

$4x = 28$ $7 + y = 10$

$x = 7$ $y = 3$

> **Key Point**
>
> Solutions to simultaneous equations always come in pairs.

Add equation 1 and equation 2 to eliminate the y terms.

Substitute your value for x into one of the equations.

Annabel buys three pears and two apples for £1.20
David buys four pears and three apples for £1.65

Work out the cost of one apple and one pear.

$3p + 2a = 120$ Equation 1

$4p + 3a = 165$ Equation 2

Equation 1 × 4: $12p + 8a = 480$

Equation 2 × 3: $12p + 9a = 495$

$a = 15$

$3p + (2 \times 15) = 120$

$p = 30$

An apple costs 15p and a pear costs 30p.

Form two equations with the information given.

Multiply so that the p terms match. Remember to multiply all terms.

Subtract equation 1 from equation 2.

Substitute your value for a into one of the equations and solve.

- When you have a non-linear equation and a linear equation, always substitute the linear into the non-linear.

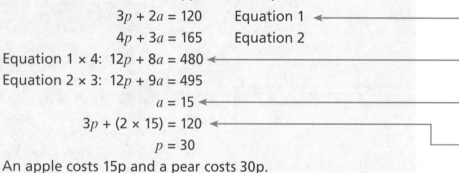

Solve the equations $2y + x = 3x^2$ and $y + 5x = -3$

$2y + x = 3x^2$ Equation 1

$y + 5x = -3$ Equation 2

$y = -3 - 5x$

$2 \times (-3 - 5x) + x = 3x^2$

$3x^2 + 9x + 6 = 0$

$x^2 + 3x + 2 = 0$

$(x + 1)(x + 2) = 0$

$x = -1, \ y = 2$ or $x = -2, \ y = 7$

Make y the subject of equation 2.

Substitute into equation 1.

This is a quadratic equation; rearrange to = 0.

Divide all terms by 3.

Solve by factorisation.

Solving Equations with Graphs

- You can plot graphs and find the point of intersection. However, the solutions will often only be approximate.

Solve the equations $y = 3x^2$ and $y + 5x = 3$

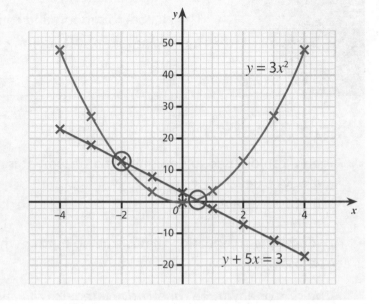

The points of intersection are (–2.1, 13.5) and (0.5, 1).

So, the two approximate solutions are $x = 0.5$, $y = 1$ or $x = -2.1$, $y = 13.5$

Functions

- **Functions** have an **input** and an **output**, e.g.

  If $f(x) = 4x + 2$ $f(5) = 4 \times 5 + 2 = 22$ 5 is the input of the function, so substitute for x.

- The inverse function is the reverse process of the function, e.g.

  If $f(x) = 4x + 2$ $f^{-1}(x) = \dfrac{x - 2}{4}$

  > **Key Point**
  >
  > The inverse function is written as $f^{-1}(x)$.

- A **composite function** is when one function becomes the input of another, e.g.

  This means make $g(x)$ the input of $f(x)$.

  $g(x) = 3x - 2$

  Expand the brackets to simplify.

  $f(x) = 2x^2 + 5$

  This means make $f(x)$ the input of $g(x)$.

  a) Work out $fg(x)$
  $$fg(x) = f(3x - 2)$$
  $$= 2(3x - 2)^2 + 5$$
  $$= 18x^2 - 24x + 13$$

  b) Work out $gf(x)$
  $$gf(x) = g(2x^2 + 5)$$
  $$= 3(2x^2 + 5) - 2$$
  $$= 6x^2 + 13$$

  Expand the brackets to simplify.

  > **Key Words**
  >
  > linear equation
  > simultaneous equations
  > function
  > input
  > output
  > composite functions

Quick Test

1. Solve the simultaneous equations $2x + y = 5$ and $x + y = 3$
2. Solve the simultaneous equations $y = 3x - 1$ and $y = x^2 - 5$
3. Solve the simultaneous equations $y = x$ and $y = x^2 - 2$ using a graph.
4. If $f(x) = 3x + 1$ and $g(x) = 2x^2 + 5$, work out $gf(x)$.

Algebraic Proof

You must be able to:

- Argue mathematically
- Use algebra to support and construct arguments
- Include proofs in your arguments.

Using Algebra

- Algebra can be used to determine if a statement is:
  - always true
  - true for certain values
  - never true.
- An **identity** is an equation that is true no matter what values are chosen.

> **Key Point**
>
> $\equiv$ means **always** equal to
>
> $\equiv$ is the identity sign and it means that any equation it is used in will always be true for any value of the variable(s).

> *The sum of any three consecutive integers is always a multiple of 3.*
>
> Is this statement true? You must show your working.
>
> Let n represent the first integer.
> The second integer is $n + 1$ and the third integer is $n + 2$.
> The sum is: $n + (n + 1) + (n + 2) \equiv 3n + 3$
> This can be written as $3(n + 1)$
> Therefore, the sum of three consecutive integers is always a multiple of 3. The statement is **always** true.

> *The sum of the squares of any three consecutive integers is always a multiple of 3.*
>
> Is this statement true? You must show your working.
>
> Let n represent the first integer.
> The second integer is $n + 1$ and the third integer is $n + 2$.
> The sum of the squares is: $n^2 + (n + 1)^2 + (n + 2)^2 \equiv 3n^2 + 6n + 5$
> 3 is **not** a factor of this expression, therefore the statement is **never** true.

Algebraic Proof

- Algebra can be used to help construct an **argument**.
- **Proofs** are examples of mathematical reasoning, used in an argument, to show a statement is true, i.e. they provide evidence for an argument.

The median of three consecutive integers is n. Show that the mean is also n.

The median is the middle number, so the three consecutive integers in terms of n are: $n - 1$, n, $n + 1$

Mean $= \dfrac{(n-1) + n + (n+1)}{3} = \dfrac{3n}{3} = n$

Show that $\dfrac{3}{x+1} + \dfrac{3x}{x^2 + 3x + 2} \equiv \dfrac{6}{x+2}$

$\dfrac{3}{x+1} + \dfrac{3x}{x^2 + 3x + 2} \equiv \dfrac{3}{x+1} + \dfrac{3x}{(x+1)(x+2)}$

Start with the LHS (left-hand side). Factorise the denominator.

$\dfrac{3}{x+1} + \dfrac{3x}{(x+1)(x+2)} \equiv \dfrac{3(x+2)}{(x+1)(x+2)} + \dfrac{3x}{(x+1)(x+2)}$

Find a common denominator.

$\dfrac{3(x+2)}{(x+1)(x+2)} + \dfrac{3x}{(x+1)(x+2)} \equiv \dfrac{3(x+2) + 3x}{(x+1)(x+2)}$

Simplify.

$\dfrac{3(x+2) + 3x}{(x+1)(x+2)} \equiv \dfrac{6x+6}{(x+1)(x+2)}$

$\dfrac{6x+6}{(x+1)(x+2)} \equiv \dfrac{6(x+1)}{(x+1)(x+2)}$

Factorise the numerator.

$\dfrac{6(x+1)}{(x+1)(x+2)} \equiv \dfrac{6}{(x+2)} =$ RHS

Cancel down.

The diagram shows two squares. The blue square has a side length of $2n$.

Show that the ratio of the area of the green square to the area of the blue square is 1 : 2

> **Key Point**
>
> Always show every stage of your working.

Area of blue square is: $2n \times 2n = 4n^2$

The side length of the green square is: $\sqrt{n^2 + n^2} = \sqrt{2n^2}$

Use Pythagoras' Theorem.

The area of the green square is: $\sqrt{2n^2} \times \sqrt{2n^2} = 2n^2$

Ratio of areas is: $2n^2 : 4n^2$, which simplifies to 1 : 2

Quick Test

1. Show that the sum of any two consecutive integers will always be an odd number.
2. Show that the difference of the squares of two consecutive integers will always be an odd number.
3. A set of five consecutive integers has a median of n. Work out the mean in terms of n.
4. Show that $\dfrac{2}{x} + \dfrac{3}{x^2} \equiv \dfrac{2x+3}{x^2}$

> **Key Words**
>
> identity
> argument
> proof

Circles

You must be able to:

- Identify and use different properties of circles
- Apply and prove theorems relating to circles
- Understand cyclic quadrilaterals.

Circle Theorems

• The angle **subtended** (formed) at the centre of a circle is twice the angle at the **circumference** if both are subtended from the same **chord** or **arc**.	• Angles subtended from the same chord or arc to the circumference are equal.	• Any angle subtended from each end of the diameter will be 90°.
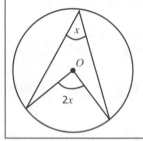		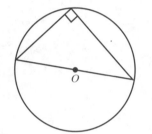

Cyclic Quadrilaterals

- All four corners of a **cyclic quadrilateral** touch the circumference of the circle.
- Opposite angles of a cyclic quadrilateral add up to 180°.

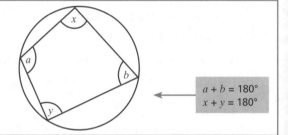

$a + b = 180°$
$x + y = 180°$

Tangents and Chords

- A **tangent** is a straight line that touches the circumference of the circle at a single point.
- If a tangent and a **radius** meet at a point on the circumference, the angle between them is 90°.

• If two tangents are drawn from the same external point to a circle, they will be of equal lengths. $AB = AC$ Triangle OAB and triangle OAC are congruent.

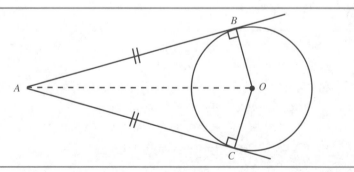

- The line joining the centre of the circle to the midpoint of a chord is perpendicular to the chord.

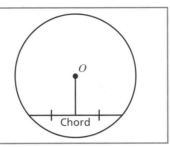

Sectors

- The **sector** of a circle is a section of the circle bounded by two radii and an arc.
- If the angle at the centre is θ then:

  Sector Area = $\dfrac{\theta}{360} \times \pi r^2$

  Arc Length of Sector = $\dfrac{\theta}{360} \times 2\pi r$

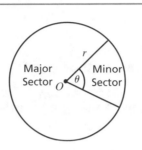

Alternate Segment Theorem

- The angle between the tangent and the chord is equal to the angle subtended by the chord in the alternate **segment**.

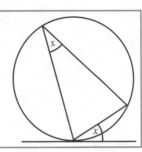

Quick Test

1. Work out the size of the lettered angle in each diagram.

a)

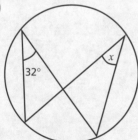

b)

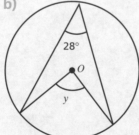

c)

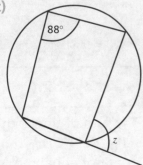

d)

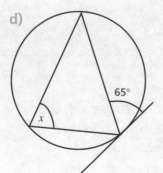

Vectors

You must be able to:

- Add and subtract vectors
- Multiply a vector by a scalar
- Work out the magnitude of a vector
- Use vectors in geometric arguments and proofs.

Properties of Vectors

- A **vector** is a quantity that has both **magnitude** (size) and direction.
- Vectors are equal only when they have equal magnitudes and are in the same direction.

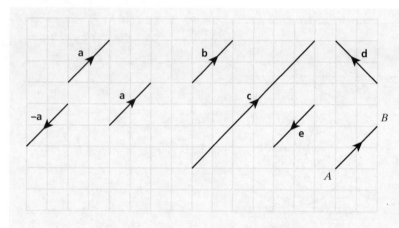

b = **a** (same direction, same length)
d ≠ **a** (different direction, same length)
e = –**a** (opposite direction, same length)
c = 3**a** (same direction, 3 × length of **a**)

$$\mathbf{a} = \overrightarrow{AB} = \underline{a} = \begin{pmatrix} 2 \\ 2 \end{pmatrix}$$

 These are all ways of writing the same vector.

$$-\mathbf{a} = \begin{pmatrix} -2 \\ -2 \end{pmatrix}$$

a and **c** are parallel vectors.
a and **b** are equal vectors.

- Any number of vectors can be added together.

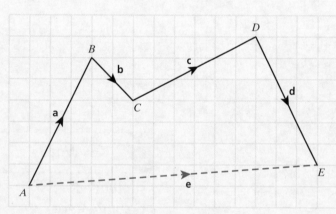

$$\mathbf{a} + \mathbf{b} + \mathbf{c} + \mathbf{d} = \begin{pmatrix} 3 \\ 6 \end{pmatrix} + \begin{pmatrix} 2 \\ -2 \end{pmatrix} + \begin{pmatrix} 6 \\ 3 \end{pmatrix} + \begin{pmatrix} 3 \\ -6 \end{pmatrix}$$

$$= \begin{pmatrix} 14 \\ 1 \end{pmatrix}$$

$\overrightarrow{AB} + \overrightarrow{BC} + \overrightarrow{CD} + \overrightarrow{DE} = \overrightarrow{AE}$ or **e** This is the resultant vector.

- When a vector is multiplied by a **scalar** (a numerical value), the resultant vector will always be parallel to the original vector.
- When a vector is multiplied by a positive number (not 1), the direction of the vector does not change, only its magnitude.

- When a vector is multiplied by a negative number (not -1), the magnitude of the vector changes and the vector points in the opposite direction.
- The magnitude of a vector **a** is written $|a|$
- Magnitude of vector $\begin{pmatrix} x \\ y \end{pmatrix}$ is $\sqrt{x^2 + y^2}$

Work out the magnitude of vector **a**.

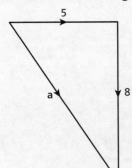

Vector $\mathbf{a} = \begin{pmatrix} 5 \\ -8 \end{pmatrix}$

$a^2 = 5^2 + 8^2$ ← Use Pythagoras' Theorem.

$a^2 = 25 + 64$

$|a| = \sqrt{89}$

$|a| = 9.43$ (to 3 significant figures)

Vectors in Geometry

- All rules of algebra can be applied to vector expressions.

In triangle DEF, G and H are the midpoints of DE and DF.

Prove that $GH = \frac{1}{2}EF$ and that GH is parallel to EF.

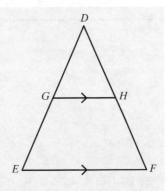

Let $\overrightarrow{DE} = \mathbf{e}$ and $\overrightarrow{DF} = \mathbf{f}$

So, $\overrightarrow{EF} = \overrightarrow{ED} + \overrightarrow{DF}$

$\qquad = -\mathbf{e} + \mathbf{f} = \mathbf{f} - \mathbf{e}$

Also $\overrightarrow{DG} = \frac{1}{2}\overrightarrow{DE} = \frac{1}{2}\mathbf{e}$

$\qquad \overrightarrow{DH} = \frac{1}{2}\overrightarrow{DF} = \frac{1}{2}\mathbf{f}$

So, $\overrightarrow{GH} = \overrightarrow{GD} + \overrightarrow{DH}$

$\qquad = -\frac{1}{2}\mathbf{e} + \frac{1}{2}\mathbf{f} = \frac{1}{2}\mathbf{f} - \frac{1}{2}\mathbf{e} = \frac{1}{2}(\mathbf{f} - \mathbf{e})$

$\overrightarrow{GH} = \frac{1}{2}\overrightarrow{EF}$

Therefore, $GH = \frac{1}{2}EF$ and GH is parallel to EF. ← This is called the midpoint theorem.

Quick Test

1. Vector **a** has a magnitude of 3cm and a direction of 120°.
   Vector **b** has a magnitude of 4cm and a direction of 040°.
   Draw the vectors:
   a) a   b) b   c) –a   d) 2a   e) a + b

Key Words

vector
magnitude
scalar

Uses of Graphs & Other Graphs 1 & 2

1 The graph below shows the journey of a car.

Estimate the total distance travelled.

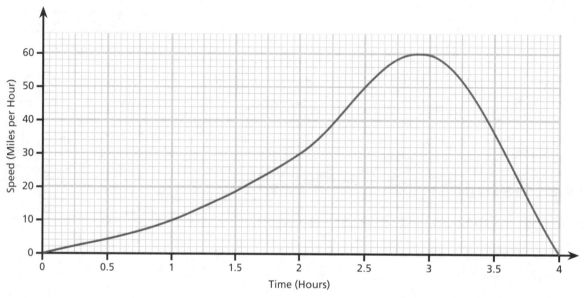

[3]

2 Line 1 has equation $5y = 3x - 2$
Line 2 has equation $3y + 5x = 6$

Show that line 1 and line 2 are perpendicular. [2]

3 The graph shows Sophie's journey to her friend's house. Her friend lives 18km away.
Sophie begins her journey at 1pm.

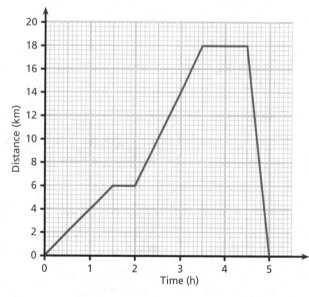

a) Sophie stops on the way to see her friend.

  i) How is this shown on the graph? [1]

  ii) How long does Sophie stop for? [1]

b) At what time does Sophie arrive at her friend's house? [1]

c) Sophie is picked up from her friend's house by her mum.

  i) Calculate the average speed, in km/h, of her journey home. [2]

  ii) At what time does Sophie arrive home? [1]

Uses of Graphs & Other Graphs 1 & 2 (Cont.)

4 The graph below has equation $y = x^2$

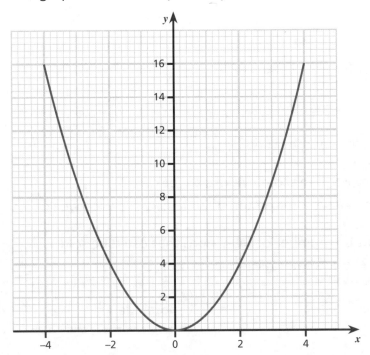

Estimate the gradient of the curve at the point (2, 4). [1]

5 A circle has centre (0, 0) and goes through the point (3, 5).

a) Work out the equation of the circle. [3]

b) Work out the equation of the tangent to the circle at the point (3, 5). [3]

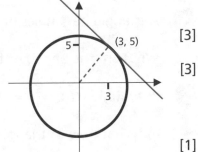

6 a) y is directly proportional to x.

Sketch a graph to show this relationship. [1]

b) y is inversely proportional to x.

Sketch a graph to show this relationship. [1]

7 a) Expand $(x - 3)(x + 3)(x - 2)$ [3]

b) Sketch the graph of $(x - 3)(x + 3)$ [2]

8 Work out the equation of the line that is perpendicular to the line with equation
$2y + x = 5$ and goes through the point (3, 6). [3]

Total Marks _____ / 28

Review Questions

Inequalities

1 Work out all the possible integer values of n if $3 \leqslant n \leqslant 7$. [1]

2 Solve $2x - 6 > 2$ [2]

3 Work out all the possible integer values for y if $12 \leqslant 3y \leqslant 36$. [2]

4 Solve $x^2 + 5x - 24 \geqslant 0$ [3]

5 Solve $2 - 9x > 3$ and show the solution on a number line. [2]

6 A TV salesperson is set a target to sell more than six televisions a week.
The manufacturer can let the salesperson have a maximum of 20 televisions each week.

Use an inequality equation to represent the number of televisions that could be sold each week if the salesperson meets or exceeds their target. [2]

Total Marks _____ / 12

Congruence and Geometrical Problems

1 A triangle has angles of 56°, 64° and 60°. The triangle is enlarged by scale factor 2.

What are the angles of the enlarged triangle? [3]

2 In triangle ABC, AB = 10cm, BC = 12cm and CA = 10cm.

a) What type of triangle is triangle ABC? [1]

b) D is a point on BC such that AD is the perpendicular bisector of BC.

Prove that triangle ABD is congruent to triangle ACD. [3]

c) Calculate the area of triangle ABC. [3]

Total Marks _____ / 10

Right-Angled Triangles

1 One of the pyramids of Egypt is built on a square base of width 200m.

C is the centre of the base. The slant height *TB* is 310m.

Work out the perpendicular height of the pyramid to the nearest metre.

[4]

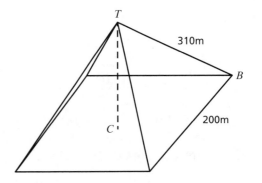

2 A rectangle measures 3cm by 4cm. 🔲

What is the length of its diagonal? [1]

3 Calculate the area of the triangle. 🔲 [3]

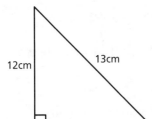

4 *PCDEF* is a pyramid with a rectangular base *CDEF*.
CD = 8cm and *DE* = 5cm.
P is vertically above the centre of the rectangle and *PC* = *PD* = *PE* = *PF* = 13cm.

Calculate:

a) The angle between *PC* and the plane *CDEF*. Give your answer to 1 decimal place. [4]

b) The vertical height of *P* above the rectangular base. Give your answer to 3 significant figures. [2]

5 Two polar bears, Snowy and Blizzard, are asleep in an enclosure.
The distance from Snowy to Blizzard is 25m. The bearing of Blizzard from Snowy is 078°.

Calculate how far east Blizzard is from Snowy. Give your answer to 2 decimal places. [2]

Total Marks _____ / 16

Sine and Cosine Rules

1 Here is a triangle ABC.

a) Which angle is the smallest?
Give a reason for your answer. [2]

b) Calculate the size of the smallest angle to the nearest degree. [4]

c) Calculate the area of triangle ABC. [3]

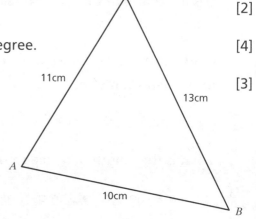

2 A fishing boat leaves harbour and sails 12km on a bearing of 040° to a lighthouse.

The boat then changes direction and sails on a bearing of 145° to visit Seal Island, which is 14km due east from the harbour.

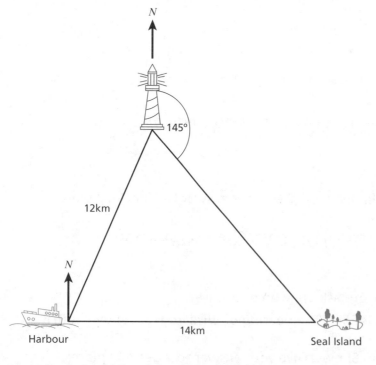

How far is Seal Island from the lighthouse?
Give your answer to 1 decimal place. [4]

Total Marks _____ / 13

Statistics 1 & 2

1 This histogram gives information about the books sold in a bookshop one Friday.

a) Use the histogram to complete the table. [2]

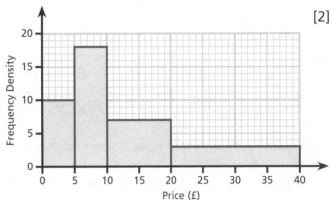

Price (pounds)	Frequency
$0 < P \leqslant 5$	
$5 < P \leqslant 10$	
$10 < P \leqslant 20$	
$20 < P \leqslant 40$	

b) Estimate the mean price of books sold. [3]

c) Work out the class interval that contains the median. [2]

2 Corinna wants to sample 100 people in Malmesbury to find out how often people visit the library. She stands by the entrance of the library on a Thursday morning and asks the first 100 people she meets.

a) Explain what is wrong with this sample. [1]

b) Describe a better method that Corinna could use to collect the data. [2]

3 The scatter diagram below shows the body temperature and pulse rate of 10 animals.

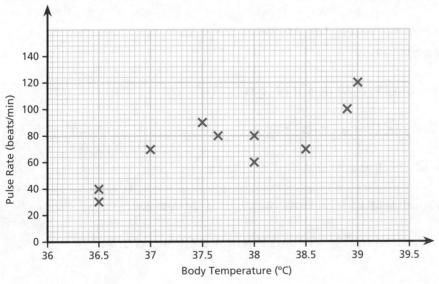

a) Draw a line of best fit and use it to estimate the pulse rate of an animal with a body temperature of 39.5°C. [2]

b) Comment on the reliability of this estimate. [1]

Total Marks .... / 13

Measures, Accuracy and Finance

1 Estimate an answer to the following calculation:

$$\frac{512 \times 7.89}{0.19}$$

[2]

2 4kg of carrots and 5kg of potatoes cost £6.11

If 5kg of carrots cost £4.45, work out the cost of 4kg of potatoes. [3]

3 Sabrina's take-home pay is worked out using this formula:

Take-Home Pay = Hours Worked × Hourly Rate – Deductions

Sabrina's hourly rate is £35.
Her deductions were £218.
Her take-home pay was £657.

Work out the number of hours she worked that week. [3]

4 Calculate $\dfrac{\sqrt{(6.2^2 - 3.6)}}{2.6 \times 0.15}$

Give your answer to **a)** 2 decimal places and **b)** 3 significant figures. [4]

5 The price of a new television is £560, including tax at 17.5%.

Work out the cost of the television before tax is added.
Give your answer to the nearest penny. [2]

6 Graham uses the following formula to work out his gas bill:

Cost = Standing Charge + Cost per Unit × Units Used

The standing charge is £50.93 and the cost per unit is 3p:

a) Last month Graham's gas bill was £53.15. How many units did he use? [2]

b) The standing charge is increased by 3%. What is the new standing charge? [2]

Total Marks .... / 18

Solving Quadratic Equations

1   Solve the equation $3x^2 = 27$ [1]

2   **a)** Write $x^2 + 8x - 12$ in the form $(x + p)^2 + q$, where p and q are integers.   [2]

     b) Solve the equation $x^2 + 8x - 12 = 0$ giving your answer in the form $a \pm b\sqrt{7}$.   [2]

     c) Write down the minimum point of $y = x^2 + 8x - 12$ [1]

3   The diagram shows a trapezium. All the measurements are in centimetres. The area of the trapezium is 14cm².

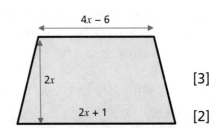

     a) Show that $6x^2 - 5x - 14 = 0$ [3]

     b) Solve the equation to find the value of x.   [2]

> **Total Marks** _____ / 11

Simultaneous Equations and Functions

1   **a)** On the same set of axes, draw the graphs $y = 2x^2$ and $y = 3x + 2$ [2]

     b) Use your graph to solve the equation $2x^2 = 3x + 2$ [2]

2   Solve the simultaneous equations: [4]

     $2x + y = 1$

     $y = x^2 - 2$

3 $f(x) = x^2 + 1$ and $g(x) = 3x - 2$

     a) Work out $fg(x)$ [2]

     b) $fg(x) = gf(x)$

     Show that this statement is incorrect.   [2]

     c) Use the quadratic formula to solve the equation $fg(x) = 10$ [3]

> **Total Marks** _____ / 15

Algebraic Proof

1 Prove that the sum of the squares of any three consecutive integers is 1 less than a multiple of 3. [3]

2 Grace surprises her friends with the following number trick:
- Think of a number.
- Multiply your number by 3.
- Add 30 to your answer.
- Divide your number by 3.
- Subtract the number you first thought of.
- Your answer is 10.

Use algebra to prove that Grace's trick always works. [3]

3 Show that $\dfrac{3}{x+2} + \dfrac{9}{x^2+x-2} = \dfrac{3}{x-1}$ [3]

4 Show that $\dfrac{6}{3x+2} + \dfrac{4}{3x-2} \equiv \dfrac{30x-4}{9x^2-4}$ [3]

5 The diagram below shows a circle inside a square. The square has a side length of $6x$.

Show that the ratio of the area of the square to the area of the circle is $4 : \pi$ [4]

6 Prove that $(4n+1)^2 - (4n-1)^2$ is a multiple of 8 for all positive integer values of n. [3]

7 Prove that the difference between the squares of any two consecutive even numbers is always a multiple of 4. [3]

8 a and b are numbers such that $a = b + 2$

The sum of a and b is equal to the product of a and b.

Show that a and b are **not** integers. [3]

Total Marks .... / 25

Circles

1 For each of the following questions, work out the lettered angles.
The centre of each circle (where appropriate) is marked with an O.

a) [1]

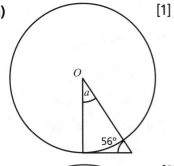

b) [2]

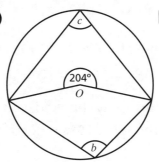

c) [2]

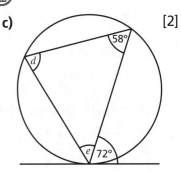

d) [2]

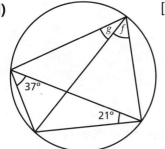

e) [2]

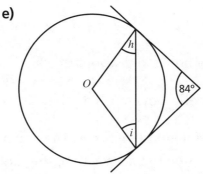

f) [1]

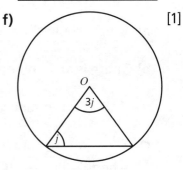

Total Marks .... / 10

Vectors

1 ABC is a triangle. D is the midpoint of AB. E is the midpoint of BC.
$\overrightarrow{AB} = \mathbf{a}$ and $\overrightarrow{AC} = \mathbf{c}$

a) Work out $\overrightarrow{AE}$ in terms of $\mathbf{a}$ and $\mathbf{c}$. [3]

b) Show that $\overrightarrow{DE}$ is parallel to $\overrightarrow{AC}$. [3]

2 State whether the following can best be described as a vector or as a scalar:

a) The length of a desk [1]

b) The flight path of an aircraft [1]

c) A snooker ball hit directly into a pocket. [1]

Total Marks .... / 9

Review Questions

Measures, Accuracy and Finance

1 Calculate the value of $\dfrac{25.75 \times 31.3}{7.62 - 1.48}$

Give your answer to **a)** 2 decimal places and **b)** 3 significant figures. [3]

2 The Cotton family, comprising of Mum, Dad and 16-year-old daughter, are going away to Turkey.

The brochure states that for every three days booked, you receive one day free. The price per adult per day is £42. There is a 15% reduction for a child up to the age of 17. The package is all inclusive.

a) What would the total cost of the holiday be if the family books 12 days? [3]

b) There is an additional early booking discount of 5%.

If the Cotton family book early, how much do they save? [2]

c) The flight to Turkey takes 4 hours 20 minutes. The departure time from Gatwick is scheduled for 11.10am but, due to bad weather, the flight is delayed by 1 hour 34 minutes.

If Turkish time is two hours ahead of UK time, at what time did the flight arrive in Turkey? [2]

3 What is the maximum area of a square of side 4.2cm measured to the nearest millimetre? [1]

4 Meka wants to buy a mobile phone. There are two different options:

Option 1: Phone costs £146 + £80 per month

Option 2: Phone costs £548 + £43 per month

Both options come with a 5% annual discount.

a) Which option is cheaper for a one-year period? [3]

b) How much cheaper is this option? [1]

Total Marks .... / 15

Solving Quadratic Equations

1 A circle has equation $x^2 + y^2 = 36$

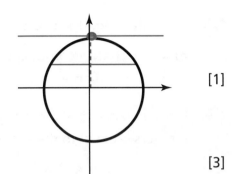

   a) Write down the equation of the tangent shown on the diagram. [1]

   b) A chord, as shown on the diagram, goes through the point (0, 3) and intersects the circle in two places. Work out the points of intersection. [3]

2 **a)** Write $x^2 - 12x + 26$ in the form $(x + p)^2 + q$. [3]

   b) Use your answer from part **a)** to solve the equation $x^2 - 12x + 26 = 0$
   Leave your answers in surd form. [2]

3 Solve the equation $\dfrac{4}{x} + \dfrac{4}{x + 2} = 3$ [4]

4 The area of the triangle is 4cm².

Work out the value of x. Give your answer in the form $a\sqrt{6}$. [3]

$\frac{x}{15}$ cm

$\frac{x}{5}$ cm

5 Use the quadratic formula to solve the equation $x^2 - 7x - 15 = 0$ [2]

Total Marks _____ / 18

Simultaneous Equations and Functions

1 **a)** On the same axes, draw the graphs of $y = 3x^2$ and $y = 4x + 2$. [2]

   b) Use your graphs to find estimates to the solutions of the equation $3x^2 = 4x + 2$ [2]

2 Solve the following simultaneous equations:

$y - x = 4$

$y = x^2 + 2$ [4]

Total Marks _____ / 8

Review Questions

Algebraic Proof

1   **a)** Write down the expression for the nth term of the sequence 4, 7, 10, 13, 16 ...   [1]

      b) Prove that the product of any two consecutive terms in this sequence is also a term of the sequence.   [3]

2   Prove algebraically that the difference between the squares of any two consecutive integers is equal to the sum of these two integers.   [3]

3   Show that $\dfrac{x^2 + 5x + 6}{x^2 + 2x} = \dfrac{x + 3}{x}$ [3]

4   Jessa thinks that when you subtract 7 from a number, the answer is always odd.

      Give an example to show that Jessa is wrong.   [1]

5   The shape shown is a square.

      Emma states that the area of the square is half of the product of the diagonals. Prove that Emma is correct.   [2]

6   Prove that $(3n + 1)^2 - (3n - 1)^2$ is a multiple of 4 for all positive integer values of n.   [3]

7   Here is a sequence of numbers:
      2, 5, 10, 17, 26, 37 ...

      a) Write down the expression for the nth term of this sequence.   [1]

      b) Show algebraically that the sum of two consecutive terms in this sequence will always be odd.   [3]

8   Show that $\dfrac{2}{x - 2} - \dfrac{8}{x^2 - 4} \equiv \dfrac{2}{x + 2}$ [3]

9   Prove that the square of any odd number is always one more than a multiple of 4.   [3]

10   *'The sum of two prime numbers is also a prime number.'*

      Give an example to disprove this statement.   [1]

Total Marks _____ / 27

Circles

1 Work out the size of the angle marked with a letter. The centre of each circle (where appropriate) is marked with an O. 📱

a) [1] **b)** [1] **c)** [2]

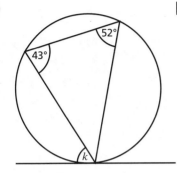

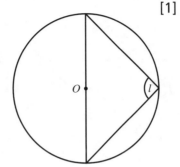

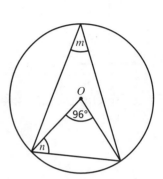

2 a) Label each arrow pointing to a different part of the circle. [5]

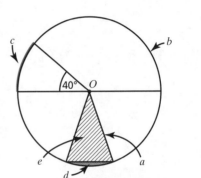

b) Calculate the length of c if the radius = 10cm (take π = 3.14).
Give your answer to 2 decimal places. [3]

Total Marks _____ / 12

Vectors

1 $ABCD$ is a parallelogram and AC is one of the diagonals.
P is a point on AC such that $AP = \frac{1}{4} AC$
$\vec{DA}$ = 4**a** and $\vec{DC}$ = 4**c**

a) Work out the vector $\vec{AC}$ in terms of **a** and **c**. [1]

b) Work out the vector $\vec{AP}$ in terms of **a** and **c**. [1]

c) Work out the vector $\vec{DP}$ in terms of **a** and **c**. [1]

2 Write down whether each of these statements is **true** or **false**.

a) Velocity is a vector. [1]

b) Two vectors are equal if they have the same magnitude but are in opposite directions. [1]

Total Marks _____ / 5

Mixed Exam-Style Questions

1 The diagram shows a regular pentagon of side length P.

Shade the area on the diagram that is represented by the expression $\frac{1}{2}CG + \frac{1}{2}PG$.

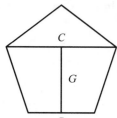

[1]

2 Factorise completely $3x^3 + 6x^2$

Answer _____ [1]

3 Solve $3(x + 3) - 5 = 2(x - 2)$

Answer _____ [3]

4 Work out the value of $\frac{5}{7}\left(x - \frac{2}{5}\right) + 9y$ when $x = \frac{1}{5}$ and $y = \frac{2}{7}$

Answer _____ [2]

5 Simplify $3x^2 + 6x - 2x^2 + 4x - 2y - 6x^2$

Answer _____ [2]

6 Simplify $3p^2y - py + p^2y + 7py$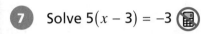

Answer _____ [2]

7 Solve $5(x - 3) = -3$

Answer _____ [2]

8 Write $(2x - 2)(x - 3)(3x + 4)$ in the form $ax^3 + bx^2 + cx + d$ where a, b, c and d are integers.

Answer _____ [2]

9 **a)** Factorise $x^2 - x - 2$

Answer _____ [1]

b) Factorise $2x^2 - 5x + 2$

Answer _____ [1]

c) Using your answers from parts **a)** and **b)** simplify $\dfrac{2x^2 - 5x + 2}{x^2 - x - 2}$

Answer _____ [2]

10 Use the formula $P = 3r - q^2$ to work out the exact value of q when $P = 30$ and $r = 20$.

Answer _____ [2]

11 **a)** Factorise $x^2 - 16$

Answer _____ [1]

b) Solve $x^2 - 16 = 0$

Answer _____ [2]

12 The formula used to calculate the area of a circle is $A = \pi r^2$.

A circle has an area of 25cm².

Ethan thinks the radius of the circle is $\dfrac{5}{\sqrt{\pi}}$

Guy thinks the radius is $\dfrac{\sqrt{\pi}}{5}$

Who is correct? You must show how you got your answer.

_____

_____ [2]

13 Expand and simplify $\sqrt{5}\left(\sqrt{5} + 3\right)$

Answer _____ [2]

14 Write the following numbers in order, smallest first.

6.77   6.767   6.677   6.8

Answer _____ [1]

15 **a)** Write 45 as a product of prime factors.

Answer _____ [2]

b) Write 105 as a product of prime factors.

Answer ⎯⎯⎯⎯⎯⎯⎯⎯⎯⎯⎯⎯ [1]

c) Use your answers to parts **a)** and **b)** to work out the highest common factor (HCF) of 45 and 105.

Answer ⎯⎯⎯⎯⎯⎯⎯⎯⎯⎯⎯⎯ [2]

d) Work out the lowest common multiple (LCM) of x^2y and y^2x, where x and y are prime numbers.

Answer ⎯⎯⎯⎯⎯⎯⎯⎯⎯⎯⎯⎯ [2]

16 Work out $5\frac{1}{6} - 2\frac{1}{3}$

Answer ⎯⎯⎯⎯⎯⎯⎯⎯⎯⎯⎯⎯ [3]

17 $P = xy$

x is increased by 10%.
y is increased by 10%.

Work out the percentage increase in P.

Answer ⎯⎯⎯⎯⎯⎯⎯⎯⎯⎯⎯⎯ [2]

18 Mandeep is looking for a new 12-month phone contract.

Dave's Dongles	Ian's Internet
£12 a month	£10 a month
+	+
5p a minute	6p a minute
10% discount on first 6 months	15% discount on first 4 months

On average Mandeep uses 120 minutes per month.

Which phone contract is cheaper for Mandeep?
You must show your working.

Answer _____ [5]

19 97 × 1452 = 140 844

Use this information to write down the value of:

a) 9.7 × 145.2

Answer _____ [1]

b) 0.97 × 0.1452

Answer _____ [1]

20 $y = \dfrac{ab}{a + b}$

$a = 3 \times 10^4$

$b = 5 \times 10^3$

Work out the value of y correct to five significant figures.

Give your answer in standard form.

Answer _____ [2]

21 Write down the formula for calculating speed. 🔢

Answer _____ [1]

22 A regular polygon has 20 sides.

Work out the size of each interior angle.

Answer _____ [2]

23 Write down the nth term of the following sequence of numbers:

8, 11, 14, 17, 20 …

Answer _____ [1]

24 A cat rescue centre recorded the age of the cats it rehomed over the course of a year.

Age	Frequency
$0 \leqslant a < 2$	256
$2 \leqslant a < 4$	219
$4 \leqslant a < 6$	165
$6 \leqslant a < 8$	120
$8 \leqslant a < 10$	65

a) Write down the group that contains the median.

Answer _____ [1]

b) Calculate an estimate of the mean age.

Answer _____ [3]

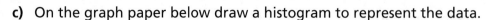

c) On the graph paper below draw a histogram to represent the data.

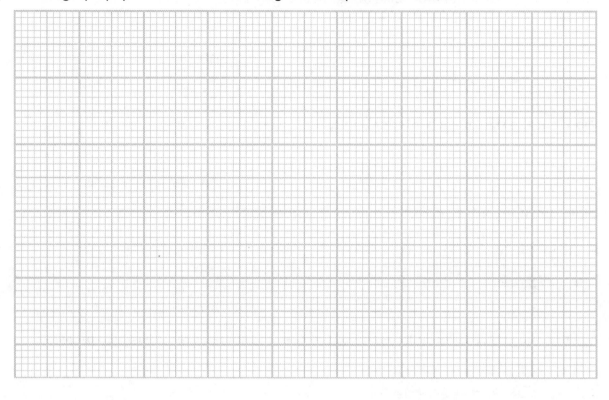

[2]

d) The rescue centre runs a campaign to help rehome the older cats and a further 78 cats over the age of 5 are rehomed.

If this information is added to the data, does it affect the group in which the median lies? Explain your answer.

_____

_____

_____ [2]

25 The force (F) between two magnets is inversely proportional to the square of the distance (x) between them.

$F = 4$ when $x = 3$

a) Work out an expression for F in terms of x.

Answer _____ [2]

b) Work out the value of F when $x = 5$.

Answer _____ [1]

c) Work out the value of x when $F = 20$.

Answer _____ [2]

26 **a)** Work out the length of side x.

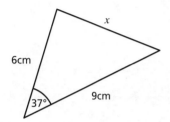

Answer _____ [3]

b) Work out the area of the triangle.

Answer _____ [2]

c) The lengths of the triangle are enlarged by scale factor 3.
The area of the new triangle can be found using the following rule:
Area of Enlarged Triangle = P × Area of the Original

Write down the value of P.

Answer _____ [1]

27 What is the equation of the y-axis?

Answer _____ [1]

28 As part of a health and safety review, a company surveys its employees to find out how many wear glasses or contact lenses.

	Male	Female
Glasses	9	6
Contact Lenses	8	16
Neither	20	15

a) Write down the ratio of the number of females who wear glasses to the number of females who wear contact lenses. Give your answer in its simplest form.

Answer _____ [1]

b) What percentage of all employees are male and do not wear glasses or contact lenses?

Answer _____ [2]

29 a) Sketch the graph of

$y = x^2 + 5x + 4$ [1]

b) Using your graph from part **a)**, solve the inequality

$x^2 + 5x \leqslant -4$

Answer _____ [2]

c) On your graph, shade the region that represents the inequality in part **b)**. [1]

30 The diagram shows a regular hexagon $ABCDEF$ with centre O.

a) $\overrightarrow{OA} = 2\mathbf{a}$ and $\overrightarrow{OB} = 2\mathbf{b}$

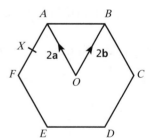

Express in terms of $\mathbf{a}$ and/or $\mathbf{b}$

i) $\overrightarrow{AB}$

Answer _____ [1]

ii) $\overrightarrow{EF}$

Answer _____ [1]

b) X is the midpoint of AF.

Express $\overrightarrow{DX}$ in terms of **a** and **b**.

Answer _____ [2]

c) Y is the point on BA extended, such that $BA : AY = 3 : 2$

Prove that D, X and Y lie on the same straight line.

Answer _____ [3]

31 Solve the simultaneous equations:

$y = x^2 - 1$

$y = 3x + 3$

Answer _____ [4]

32 In the diagram, A, B and C are points on the circumference of a circle, centre O.
Angle $BCE = 57°$

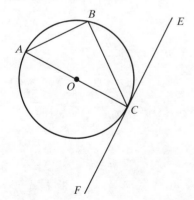

FE is a tangent to the circle at point C.

a) Calculate the size of angle ACB. Give a reason for your answer.

_____ [2]

b) Calculate the size of angle BAC.
   Give reasons for your answer.

_____ [2]

33 Simplify $(5t^{-2})^{-1}$

Answer _____ [2]

34 The diagram shows a sector of a circle, centre O.
The radius of the circle is 6cm.
The angle of the sector at the centre of the circle is 115°.

Work out the perimeter of the sector.

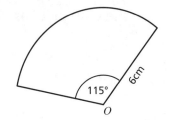

Answer _____ [4]

35 A bag contains six blue beads and five red beads.
Samuel takes a bead at random from the bag. He records its colour and replaces it.
He does this one more time.

Work out the probability that he takes one bead of each colour from the bag.

Answer _____ [3]

36 $P = 1.5$ rounded to 1 decimal place.

$Q = 2.65$ rounded to 2 decimal places.

a) Work out the maximum value of PQ.

Answer _____ [2]

b) Work out the minimum value of $\dfrac{P}{Q}$

Answer _____ [2]

37 A sketch of the graph $y = x^2 + 3x + 2$ is shown below.

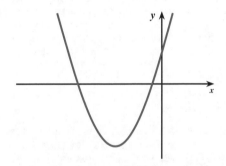

a) Work out the coordinates of the roots of the graph.

Answer _____ [2]

b) Work out the coordinates of the turning point.

Answer _____ [3]

c) Work out the coordinates of the turning point of the curve $y = f(x + 2)$

Answer _____ [1]

Total Marks _____ / 115

Answers

1. 2 000 005 [1]
2. 45kg [1]
3. $2^2 \times 3 \times 5$ [1]
4. a) 18 [1]
   b) 180 [1]
5. 8 [1]
6. –8 [1]
7. 64 [1]
8. $3 + 2m$ [1]
9. $3y^2$ [1]
10. a) 17, 20 [1]
    b) Yes [1]; nth term $= 3n + 2$ and
    $(3 \times 46) + 2 = 140$ [1]
    OR $\frac{(140 - 5)}{3} = 45$
    (term-to-term rule is +3) [1]
11. 0.06 [1]
12. 9 sides [1]
13. 3.46 [1]
14. (4.5, 7) [2]
15. 27 [1]
16. 10km/h [1]
17. 125cm³ [1]
18. 0.4 [1]
19. 0.875 [1]
20. $\frac{78}{100}$ [1]; $\frac{39}{50}$ [1]
21. $\frac{2}{3} = \frac{8}{12} = \frac{16}{24}$ and $\frac{3}{4} = \frac{9}{12} = \frac{18}{24}$ [1]; $\frac{17}{24}$ [1]
22. 10, 15, 20 [3]
23. 999 999 [1]
24. £21 : £35 [2]
25. 0.04, 0.39, 0.394, 0.4 [1]
26. a) $m = 4$ [1]
    b) $m = 48$ [1]
    c) $m = 1.5$ [1]
27. 36cm² [1]
28. 404 [1]
29. £12.75 [1]
30. $13a - 10y$ [1]
31. £81 [1]
32. $A = \pi \times r \times r$ [1]; $A = 314$cm² [1]
33. Vol (1) $= 2 \times 2 \times 2 = 8$cm³ [1];
    Vol (2) $= 4 \times 4 \times 4 = 64$cm³ [1];
    $\frac{64}{8} = 8$ [1]
34. 15 [1]
35. $\frac{63}{3} \times 4 = 84$ [2]
36. $m = -3$ [1]
37. (5, 0) [2]
38. 13cm [1]
39. $(2 \times 3^2) + (3 \times 4) = 18 + 12$ [1]; $= 30$ [1]
40. 6.5 [1]

Page 9 Quick Test
1. -0.125, 0.0125, $\frac{1}{25}$, 1.25
2. a) 24
   b) 1
   c) –12.5
   d) 1
3. a) 1.63×10^{-3}
   b) 1.63×10^7

Page 11 Quick Test
1. $2^2 \times 3^2$

2. a) 3
   b) 60
3. 10am

Page 13 Quick Test
1. $6y - 5$
2. –171
3. $12t^2 - 3t$
4. $2r^2(2r - 1)$
5. $t = -0.25$ OR $-\frac{1}{4}$

Page 15 Quick Test
1. $w = 9$
2. $y^2 + 2y - 8$
3. $(2q + 1)(q + 3)$
4. $y = \frac{x + 6}{5}$

Page 17 Quick Test
1. £108
2. 4 : 1
3. a) Box A, 1 sachet costs 31.7p; Box B,
   1 sachet costs 31.1p; Box B is the
   best buy
   b) 2

Page 19 Quick Test
1. 402mph
2. £51.11
3. 21%

Page 21 Quick Test
1. a) Rectangle, Parallelogram, Kite
   b) Square, Rhombus
2. Angle $EJH = 16°$

Page 23 Quick Test
1. a) 3240°
   b) 162°
2. 12
3.

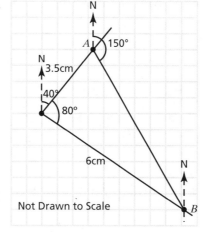

Not Drawn to Scale

Bearing $= 150° \pm 2°$

Page 25 Quick Test
1. $\frac{3}{4} \div \frac{1}{5}$ $\left(\frac{3}{4} + \frac{1}{5} = \frac{15}{20} + \frac{4}{20} = \frac{19}{20}\right.$ and
   $\left.\frac{3}{4} \div \frac{1}{5} = \frac{15}{4} = 3\frac{3}{4}\right)$
2. a) $0.\dot{1}$
   b) recurring
3. $\frac{3}{80}$

Page 27 Quick Test
1. 275 carrots
2. £10.50 + £60 = £70.50
3. 1.5%

Page 29 Quick Test
1. a) 0.1
   b) 45

Page 31 Quick Test
1. $\frac{1}{18}$
2. $\frac{5}{18}$

$$\frac{3}{9} \times \frac{2}{8} + \frac{4}{9} \times \frac{3}{8} + \frac{2}{9} \times \frac{1}{8}$$

Page 32
1. -3.38, -3, $-\frac{1}{8}$, 3.3 [2] (1 mark if one
   mistake; 0 marks if more than one
   mistake)
2. $-4 \times -4 \times -4 = -64$ [1];
   $(2^3 - 3^3) - (-7)^2 = -19 - 49 = -68$ [1];
   $-4 \times -4 \times -4$ is larger [1]
3. a) 3.84×10^5 [2] (1 mark for correct
   digits but decimal point in
   wrong position)
   b) $384\,000 - 19\,600$ [1]; $= 364\,400$
   OR 3.644×10^5 [1]

1. 84 [1]
2. a) 60 [1]
   b) 3 [1]
3. Fifth square number = 25, third
   cube number = 27 [1]; The third
   cube number is greater [1]
4. $256 = 2^8$ [1]; $n = 8$ [1]
5. No [1]; $2^3 = 8$, $3^2 = 9$ [1]
6. 2 packs of sausages [1];
   3 packs of bread rolls [1]

Page 33
1. $9x + 4y$ [2]
2. $4y - 8z + 12$ [2]
3. $4x^2 + 2x + 4$ [2]
4. 23 [1]
5. $8b = 14$ OR $2b = \frac{7}{2}$ [1]; $b = \frac{7}{4}$ OR $1\frac{3}{4}$ [1]
6. $3p + 6 = 2p + 6$ [1]; $p = 0$ [1]
7. $\frac{5}{2}x - \frac{2}{3}x = \frac{1}{2} + \frac{1}{3}$ [1]; $\frac{15}{6}x - \frac{4}{6}x = \frac{3}{6} + \frac{2}{6}$ [1];
   $x = \frac{5}{11}$ [1]

Look for a common denominator.

8. $6x - 30y + 36$ [2] (1 mark for 2 correct
   terms)
9. $6p - 4q + 12$ [2]

$- \times - = +$

10. $4xz(y - 1)$ [2] OR $4z(yx - x)$ [1] OR
    $4x(yz - z)$ [1]
11. $(w^2 + 5w + 4)(w - 4)$ [1];
    $w^3 + 5w^2 + 4w - 4w^2 - 20w - 16$ [1];
    $w^3 + w^2 - 16w - 16$ [1]
12. $(x + 1)(x + 2)$ [2]

13. $6x - 15 + 4x + 12 - 4x$ **[1]**; $6x - 3$ **[1]**; $3(2x - 1)$ **[1]**

14. $4xy + 2y = 1 - 3x$ **[1]**;
$y(4x + 2) = 1 - 3x$ **[1]**;
$y = \frac{1 - 3x}{4x + 2}$ **[1]**

15. a) $2A = (a + b)h$ **[1]**; $h = \frac{2A}{a + b}$ **[1]**
   b) $h = \frac{48}{5 + 7}$ **[1]**; 4cm **[1]**

16. $(x + 3)(x + 2)$ **[2]**

17. a) $V = \sqrt{50}$ **[1]**; $= 5\sqrt{2}$ OR 7.07 **[1]**
   b) $V^2 = u^2 - 10p$ **[1]**; $u = \sqrt{V^2 + 10p}$ **[1]**

Page 34
1. 1 : 2000 **[1]**
2. $180° \div 9 = 20°$ **[1]**; largest angle = 80° **[1]**
3. 4 people = 6 celery sticks, 1 person = $6 \div 4 = 1.5$ sticks (OR 3.5 × 6) **[1]**; 14 people = 14 × 1.5 = 21 sticks **[1]**
4. **a)** 1 person takes 24 days (inverse proportion) **[1]**; 8 people take $24 \div 8 = 3$ days **[1]**
   b) $24 \div 2 = 12$ people **[1]**

1. Density = mass ÷ volume **[1]**; Density = 4560 ÷ 400 = 11.4g/cm³ **[1]**
2. Speed = distance ÷ time **[1]**; Speed = 200 ÷ 22 = 9.09m/s **[1]**
3. Speed = distance ÷ time **[1]**; Speed = 15 ÷ 3 = 5km/h **[1]**
4. Final amount = Original amount × $\left(1 + \frac{Rate}{100}\right)^{Time}$ **[1]**; Final Amount = 4000 × $\left(1 + \frac{4}{100}\right)^4$ = 4000 × 1.04⁴ = £4679.43 **[1]**; Compound Interest = £4679.43 – £4000 = £679.43 **[1]**

Page 35
1.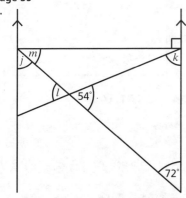
(Marks will not be awarded if reason is incorrect.) j = 72° (alternate angle) **[1]**; k = 54° (sum of the interior angles of a triangle = 180°) **[1]**; l = 54°(vertically opposite angles are equal) **[1]**; m = 18° (90° – 72°) **[1]**
2. **a)** Angle DCB = Angle DAB = 110° (opposite angles in a parallelogram are equal) **[1]**
   b) Angle ABC = 70° (allied angle to Angle BAD) **[1]**
3. $9x = 360°$, $x = 40°$ **[1]**; largest angle = 120° **[1]**
4. Exterior angle = $\frac{360}{n} = \frac{360}{10} = 36°$ **[1]**; interior angle = 180° – 36° = 144° **[1]**
5. Bearing = 180° + 36° = 216° **[1]**

6. a) Number of sides = 360° ÷ exterior angle = 360 ÷ 45 = 8 **[1]**
   b) Octagon **[1]**

Page 36
1. 32 ÷ 8 = 4 **[1]**; 32 – 4 = 28 books **[1]**
2. $\frac{1}{3} + \frac{1}{6} + \frac{1}{4} = \frac{4}{12} + \frac{2}{12} + \frac{3}{12} = \frac{9}{12}$ **[1]**; fraction that are horses is $1 - \frac{9}{12} = \frac{3}{12} = \frac{1}{4}$ **[1]**
3. $\frac{1}{6}, \frac{4}{12}, \frac{12}{24}, \frac{2}{3}$ **[2]**
4. **a)** $\frac{4}{5} \times \frac{2}{3} = \frac{8}{15}$m² **[1]**
   b) $\frac{4}{5} + \frac{4}{5} + \frac{2}{3} + \frac{2}{3} = \frac{44}{15}$m **[2]** (1 mark if answer is given as a correct mixed number)
5. d = 0.18888... , 10d = 1.8888... , 100d = 18.8888... so 100d – 10d = 17, 90d = 17 **[1]**; $d = \frac{17}{90}$ **[1]**

1. £24000 – £6000 = £18000 **[1]**; $\frac{22}{100} \times 18000 = £3960$ **[1]**
2. $\frac{46}{7200} \times 100 = 0.639\%$ **[1]**
3. £4800 × $\frac{100}{80}$ **[1]**; = £6000 **[1]**

Page 37
1. **a)** Fully correct table **[2]** (1 mark for a list of 36 outcomes)

	1	2	3	4	5	6
1	1	2	3	4	5	6
2	2	4	6	8	10	12
3	3	6	9	12	15	18
4	4	8	12	16	20	24
5	5	10	15	20	25	30
6	6	12	18	24	30	36

   b) $\frac{13}{36}$ **[1]**
   c) $\frac{8}{36} = \frac{2}{9}$ **[1]**
2. **a)** $\frac{1}{2}$ **[1]**
   b) $\frac{1}{2}$ **[1]**
3. **a)** $\frac{305}{500} = \frac{61}{100}$ **[1]**
   b) $\frac{97}{500} \times 100 = 19.4$ **[1]**; 19 or 20 **[1]**
4. **a)**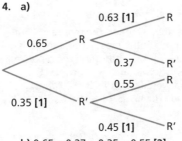
   b) 0.65 × 0.37 + 0.35 × 0.55 **[2]**; = 0.433 **[1]**
5. **a)**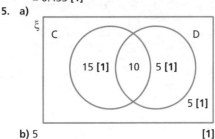
   b) 5 **[1]**

Page 39 Quick Test
1. –4, –8
2. –2, 43
3. 72

Page 41 Quick Test
1. **a)** $3n + 4$
   b) 154
2. 31, 44
3. $n^3 - 1$

Page 43 Quick Test
1. **a)** A translation by the vector $\begin{pmatrix} 5 \\ -1 \end{pmatrix}$
   b) A reflection in the line $y = x$
   c) A rotation of 90° anticlockwise about (0, 0)

Page 45 Quick Test
1. Draw a line and mark on it two points, A and B.
   Open compasses to length AB.
   Put compass point on A and draw an arc. Put compass point on B and draw an arc.
   Draw a line to join A to the new point, C.
   Adjust compasses so less than length AB.
   Put compass point on A and draw arcs crossing AB and AC at points D and E.
   Put compass point on D and draw an arc. Put compass point on E and draw an arc. Draw a line from A to the new point, F.
2.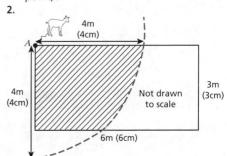

Page 47 Quick Test
1.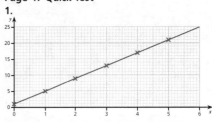
2. Gradient = –2 and y-intercept = 5
3. $y = -1.5x + 14.5$ OR $2y + 3x - 29 = 0$

Answers

Page 49 Quick Test

1.

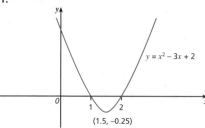

$y = x^2 - 3x + 2$

(1.5, −0.25)

2.

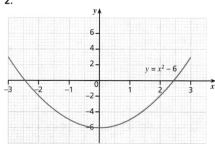

$y = x^2 - 6$

3. a) (5, 2)
   b) (8, 3)

Page 51 Quick Test

1. $6x^5$
2. 125
3. $\sqrt{2}$

Page 53 Quick Test

1. Volume = 301.59cm³ (to 2 d.p.) and surface area = 251.33cm² (to 2 d.p.)
2. 24cm²
3. Circumference = 21.99cm (to 2 d.p.) and area = 38.48cm² (to 2 d.p.)

Page 55 Quick Test

1. $\frac{500\pi}{3}$ or 523.6cm³
2. 91.5cm²
3. $\frac{200}{3}$ or 66.7cm³
4. 108π or 339.3cm²

Pages 56–61 Review Questions

Page 56

1. 12 adult tickets = £463.20 and 5 child tickets = £63.80 **[1]**; £527 **[1]**
2. a) 3.119 61 **[1]**
   b) 0.031 196 1 **[1]**
3. −243 **[1]**

1. Numbers are n, $n + 1$, $n + 2$ **[1]**; $3n + 3 = 111$, so $n = 36$ **[1]**; 36, 37, 38 **[1]**
2. a) 2 OR 9 **[1]**
   b) 35 **[1]**
   c) 9 OR 100 **[1]**
   d) 27 **[1]**
   e) 2 OR 13 **[1]**
3. 64 **[1]**
4. $25^2 \times 25^3 = 25^5$ OR $625 \times 15\,625$ = 9 765 625 = 25^5 **[1]**; $m = 5$ **[1]**
5. Peter is right **[1]**; $(2^3)^2 = 2^6 = 64$, $(2^2)^3 = 2^6 = 64$ **[1]**

Page 57

1. $k - g$ **[1]**
2. $12x - 5y$ **[2]**
3. $\frac{7}{9}$ OR $0.\dot{7}$ **[1]**
4. $(2y^2 - y - 21)(6 - y)$ **[1]**; $12y^2 - 6y - 126 - 2y^3 + y^2 + 21y$ **[1]**; $-2y^3 + 13y^2 + 15y - 126$ **[1]**
5. $b(5a - 3bc)$ **[1]**
6. 5 and 7 are prime numbers, so there are no common factors. **[1]**
7. $x = -6$ **[1]**
8. a) He has only found half of the perimeter. **[1]**
   b) $18x + 14 = 56$ **[1]**; $x = \frac{7}{3} = 2\frac{1}{3}$cm **[1]**
9. a) $\pi \times 2^2 \times 10 = 125.66$cm³ **[1]**
   b) $r^2 = \frac{V}{\pi h}$ **[1]**; $r = \sqrt{\frac{V}{\pi h}}$ **[1]**
   c) $r = \sqrt{\frac{50}{\pi \times 10}}$ **[1]**; $r = 1.26$ (to 3 significant figures) **[1]**

Page 58

1. v^2 is proportional to h so $v^2 = kh$ **[1]**; $10^2 = k \times 5$, $100 = 5k$, $k = 20$ **[1]**; $v^2 = kh$, $30^2 = 20h$, $900 = 20h$ **[1]**; $h = 45$ metres **[1]**
2. 1 part = £60 ÷ 12 = £5 **[1]**; Shares are (5 × £5) = £25, (7 × £5) = £35 **[1]**; Difference = £10 **[1]**
3. 6.2 hours = 6 hours 12 minutes = 372 minutes **[1]**; 93 : 1 **[1]**

1. Type A: Simple Interest = $\frac{\text{Amount} \times \text{Rate} \times \text{Time}}{100} = \frac{400 \times 6 \times 4}{100}$ = £96, £400 + £96 = £496 **[1]**; Type B: Final Amount = Original Amount × $\left(1 + \frac{\text{Rate}}{100}\right)^{\text{Time}}$ **[1]**; $400 \times \left(1 + \frac{5}{100}\right)^4 = 400 \times (1 + 0.05)^4$ = £486.20 **[1]**; Savings Type A gives the best return after 4 years **[1]**
2. a) Distance = Speed × Time = 1.5 × 1.5 **[1]**; = 2.25m **[1]**
   b) (11.25 × 8) ÷ 5 = 18km **[1]**; 18km/h = $\frac{18\,000\text{m}}{60\text{ minutes}}$ = 300 metres/minute = 5m/s **[1]**; After 3 seconds, the mouse is 4.5m through the pipe. After 3 seconds, the cat is 5m through the pipe. Yes, Misty will catch the mouse. **[1]**
3. Food would last (15 × 3 =) 45 days for 1 dog **[1]**; (45 ÷ 5 =) 9 days for 5 dogs **[1]**

Page 59

1. $y + 2y + 3y = 180°$, $6y = 180°$, $y = 30°$ **[1]**; largest angle = 90° **[1]**
2. 80° + 160° + 60° + fourth angle = 360° **[1]**; angle = 60° **[1]**
3. Angles around a point add up to 360°. These angles add up to 364°, so diagram is incorrect. **[1]**
4. Bearing = 180° + 054° = 234° **[1]**
5. Exterior angle = 360° ÷ n **[1]**; 360° ÷ 15 = 24° **[1]**

6. Sum of interior angles (hexagon) = 4 × 180° = 720° **[1]**; 24h = 720°, $h = 30°$ **[1]**; smallest angle (2h) = 2 × 30° = 60° **[1]**
7. a) True **[1]**
   b) True **[1]**
   c) False **[1]**

Page 60

1. $\frac{3}{5} \times \frac{2}{5} = \frac{6}{25}$ **[1]**; $\frac{4}{25} + \frac{6}{25} = \frac{10}{25} = \frac{2}{5}$ **[1]**

   Remember to use BIDMAS.

2. 36 nails **[1]**
3. $\frac{1}{16}$ **[1]**
4. $\frac{1}{8} \times 344 = 43$ **[1]**; 344 − 43 = 301 biscuits **[1]**

1. $\frac{35}{100} \times £56 = £19.60$ **[1]**; £56 − £19.60 = £36.40 **[1]**
2. Richard is correct **[1]**; $\frac{30}{100} \times £40 = £12$, $\frac{40}{100} \times £30 = £12$ **[1]**
3. $\frac{85\,000}{215\,000} \times 100 = 39.5\%$ **[1]**
4. $\frac{80}{65} \times 100$ **[1]**; = £123 **[1]**

Page 61

1. a) 0.3 **[1]**
   b) 50 × 0.3 **[1]**; = 15 **[1]**
   c) No **[1]**; P(Georgia wins) = 0.6, so greater than 0.5, which would be fair. **[1]**
2. $3x - 2 + x + 10x + 4 + 2x = 82$ **[1]**; $16x = 80$, $x = 5$ **[1]**; $2x = 10$ customers **[1]**
3. a) Venn diagram with no crossover **[1]**; All labels correct **[1]**

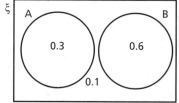

   b) 0 **[1]**
   c) 0.3 + 0.6 **[1]**; = 0.9 **[1]**

   Remember, mutually exclusive means P(A and B) = 0

4. P(vanilla and banana) = $\frac{4}{10} \times \frac{6}{9} = \frac{24}{90} = \frac{4}{15}$ **[1]**; P(banana and vanilla) = $\frac{6}{10} \times \frac{4}{9} = \frac{4}{15}$ **[1]**; $\frac{4}{15} + \frac{4}{15}$ **[1]**; = $\frac{8}{15}$ **[1]**

Pages 62–65 Practice Questions

Page 62

1. $24 - 4n$ **[2]**
2. −2, 1, 4, 7, 10 **[2]** (1 mark for any three correct)
3. a) $3n + 11$ **[2]**
   b) 311 **[1]**
4. a) $P = 10 \times 2^0$ **[1]**; $P = 10$ **[1]**
   b) $P = 10 \times 2^5$ **[1]**; $P = 320$ **[1]**

c) $2^t = 100$ **[1]**; $t = 7$ hours **[1]**
d) No, as the population will have
   a limit. **[1]**
5. a) 30, 48 **[2]**
   b) 3 **[1]**
6. a)

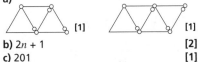

[1] **[1]**

   b) $2n + 1$ **[2]**
   c) 201 **[1]**
7.

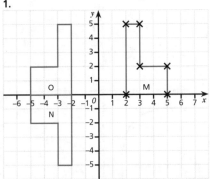

n	0	1	2	3	4	5
c	−1	4	15	32	55	84
$a + b$	5	11	17	23	29	
$2a$		6	6	6	6	

A complete table or equivalent
working **[1]**; $a = 3$, $b = 2$, $c = -1$ **[1]**;
$3n^2 + 2n - 1$ **[1]**

Page 63
1.

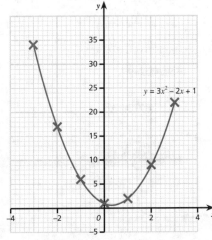

a) Shape M plotted correctly **[1]**
b) Shape N plotted correctly **[1]**
c) Shape O plotted correctly **[1]**
d) Reflection **[1]**; in the y-axis OR
   mirror line $x = 0$ **[1]**
2. a) Rectangle T is 9cm × 15cm **[1]**;
   Area = 135cm² **[1]**
   b) Area R = 15cm², Area T = 135cm² **[1]**;
   T is 9 times bigger. **[1]**
3. Draw a line and construct the
   perpendicular bisector of the line. **[1]**;
   Bisect the right angle. **[1]**
4. a) A circle **[1]**
   b) An arc of a circle **[1]**
   c) A circle **[1]**
   d) An arc of a circle **[1]**
5. a) Front elevation

[1]

   b) Plan view

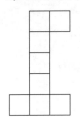

[1]

Page 64
1. Fully correct graph with y-intercept
   at (0, −2) **[1]**; and a straight line
   crossing through points (−4, −18)
   and (4, 14) **[1]**

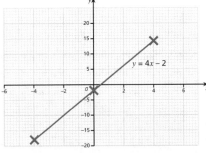

2. Fully correct graph with y-intercept at
   (0, 3) **[1]**; and a straight line crossing
   through points (−2,−7) and (2,13) **[1]**

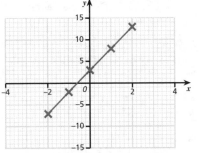

3. Gradient = −2 and y-intercept = (0, 5) **[1]**
4. Fully correct table **[1]**; and accurately
   plotted graph **[1]**

x	−3	−2	−1	0	1	2	3
y	34	17	6	1	2	9	22

5. $m = \dfrac{-1 - 5}{3 - (-2)} = \dfrac{-6}{5}$ **[1]**;

   $5 = \dfrac{-6}{5} \times \dfrac{-2}{1} + c$, $c = \dfrac{13}{5}$ **[1]**;

   $y = \dfrac{-6}{5}x + \dfrac{13}{5}$ OR $5y = -6x + 13$ **[1]**
6. $m = -1$ **[1]**; $c = 5$ **[1]**; $y = 5 - x$
   OR $x + y = 5$ **[1]**

7. a) $m = \dfrac{R - (R - 5)}{H - (H + 2)}$ **[1]** $= \dfrac{5}{-2}$ **[1]**

   b) $c = R + \dfrac{5}{2}H$ **[1]**

8. a) Minimum (−2, −8) **[1]**;
      maximum (−4, −4) **[1]**
   b) Minimum (0, −8) **[1]**;
      maximum (2, −4) **[1]**
   c) Minimum (0, −4) **[1]**;
      maximum (−2, 0) **[1]**
   d) Minimum (5, −8) **[1]**;
      maximum (3, −4) **[1]**

Page 65

1. a) 5^5 **[1]**
   b) 5^4 **[1]**
   c) $\dfrac{5^6}{5^3}$ **[1]**; $= 5^3$ **[1]**
2. $\dfrac{1}{4}$ (1 mark if 4 seen) **[2]**
3. x^8 **[1]**
4. $4 + 6\sqrt{3} + 6$ **[2]**; $= 10 + 6\sqrt{3}$ **[1]**
5. $27r^6p^3$ **[2]**

1. Area of square = 36cm² **[1]**; area of
   circles = $2 \times \pi \times 1.5^2 = 14.137...$ **[1]**;
   shaded region = $36 - 14.137...$ **[1]**;
   = 21.9cm² (to 3 significant figures) **[1]**
2. a) Volume of large cylinder =
      $\dfrac{3}{4} \times 6000\pi = 4500\pi$ **[1]**;
      $4500\pi = \pi \times 15^2 \times h$ **[1]**;
      $h = 20$cm **[1]**

   b) Volume of small cylinder =
      $\dfrac{1}{4} \times 6000\pi = 1500\pi$ **[1]**;
      $1500\pi = \pi \times r^2 \times h = \pi \times r^3$ **[1]**;
      $r = 11.4$cm (to 3 significant figures) **[1]**
3. Volume of cone = $\dfrac{1}{3} \times \pi \times 3^2 \times 7 = 21\pi$ **[1]**;
   volume of the hemisphere =
   $\dfrac{1}{2} \times \dfrac{4}{3} \times \pi \times 3^3 = 18\pi$ **[1]**; volume of
   plastic needed = $21\pi + 18\pi = 39\pi$cm³ **[1]**

Pages 66–83 Revise Questions

Page 67 Quick Test
1. $y = -2x + 15$
2. $y = 0.5x - 1.5$ OR $2y = x - 3$

Page 69 Quick Test

1.

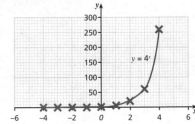

Answers

2.

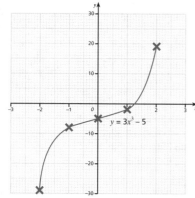

3. a) 75 miles
   b) 30 minutes

Page 71 Quick Test
1. $y = -\frac{2}{5}x + \frac{29}{5}$ OR $5y = 29 - 2x$
2. 12 (approximately)

Page 73 Quick Test
1.

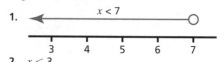

2. $x < 3$
3. Solutions are $x \geqslant 2$ and $x \leqslant -7$

Page 75 Quick Test
1. a) Angle ABC = Angle ADE
   (corresponding angles are equal); Angle
   ACB = Angle AED (corresponding
   angles are equal); Angle DAE is
   common to both triangles; so triangles
   are similar (three matching angles).

   b) $\frac{5}{10} = \frac{BC}{8}$ so BC = 4cm

Page 77 Quick Test
1. 42°
2. Yes, because $8^2 + 15^2 = 17^2$
3. 6.71cm

Page 79 Quick Test
1. a) $\frac{4}{\sin 44} = \frac{AB}{\sin 64}$; AB = 5.175cm

   b) Angle B = 72°, so area = 9.844cm²

Page 81 Quick Test
1. a) 7.76
   b) 8
   c) 9
   d) 3
2. Advantage: cheaper / quicker;
   disadvantage: could be bias / results not
   based on whole population

Page 83 Quick Test
1.

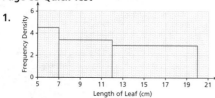

(frequency densities: 4.5; 3.4; 2.9)

---

2. a) 11cm
   b) 2.5cm

Pages 84–87 Review Questions

Page 84
1. LCM of 3 and 4 = 12 **[1]**; therefore, 29
   will be common to both **[1]**
2. a)

Pattern Number	Perimeter (cm)
1	5
2	8
3	11
4	14
60	182 **[1]**
n	$3n + 2$ **[1]**

   b) $3n + 2 < 1500$ **[1]**; 499 pentagons **[1]**
3. Yes, Jenny is correct. For n^2, even
   multiplied by even is even and odd
   multiplied by odd is odd. **[1]**; For $n^2 + 6$,
   even + 6 = even and odd + 6 = odd **[1]**
4. $2^n + 2$ **[1]**

Page 85
1. a) X = (–5, 1) **[1]**; Y = (–3, 5) **[1]**;
      Z = (–1, 2) **[1]**
   b) X = (1, –5) **[1]**;
      Y = (5, –3) **[1]**; Z = (2, –1) **[1]**
2. Lengths of rectangle D: (3 × 3 =) 9cm
   and (5.5 × 3 =) 16.5cm **[1]**; area of
   rectangle C = 16.5cm², area of rectangle
   D = 148.5cm² **[1]**; ratio is 1 : 9 **[1]**
3. a) A vertical line   **[1]**
   b) An arc of a circle   **[1]**
   c) A horizontal straight line   **[1]**
   d) An arc of a circle   **[1]**
4. A square 4cm × 4cm   **[1]**
5. Diameter of lid = 7cm **[1]**; diagram of a
   rectangle 7cm × 8cm **[1]**
6. a)   **[2]**

   b)   **[2]**

   c)   **[2]**

Page 86
1. $m = \frac{5-8}{\frac{5}{6}-\frac{2}{3}} = -18$ **[1]**;

   $8 = -18 \times \left(\frac{2}{3}\right) + c$, $c = 20$ **[1]**;

   $y = -18x + 20$ **[1]**
2. $m = \frac{4}{2} = 2$ **[1]**;

   y-intercept = (0, 3) **[1]**;

   $y = 2x + 3$ **[1]**

---

3. $(x + 1)(x + 3)$**[1]**; $x = -1$ and $x = -3$ **[1]**;

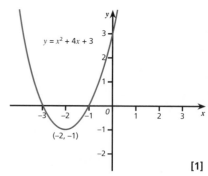

   [1]
4. $m = \frac{3}{4}$ **[1]**; $\left(0, \frac{1}{4}\right)$**[1]**
5. a) $y = (x + 7)(x - 1)$ **[1]**; $x^2 + 6x - 7$ **[1]**;
      $a = 6, b = -7$ **[1]**
   b) $y = (x + 3)^2 - 16$**[1]**; $(-3, -16)$ **[1]**
6. a)

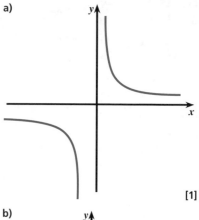

   [1]

   b)

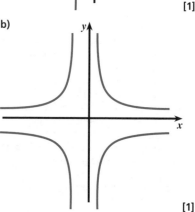

   [1]

   c) Any two from: Reflection in y-axis;
      Reflection in x-axis; Rotation 90°
      clockwise about (0, 0); Rotation 90°
      anticlockwise about (0, 0).   **[2]**

Page 87
1. $\frac{1}{8}a^{-3}b^{15}$ **[1]**; $9a^{-4}b^6$ **[1]**; $\frac{9}{8}a^{-7}b^{21}$ **[1]**
2. $\frac{2}{\sqrt{3}} - \frac{6}{3\sqrt{3}} + 6 \times 4\sqrt{3}$ **[1]**;

   $\frac{2}{\sqrt{3}} - \frac{2}{\sqrt{3}} + 24\sqrt{3}$ **[1]**; $24\sqrt{3}$ **[1]**
3. True **[1]**; $x^{-6} = \frac{1}{x^6}$ **[1]**
4. $\frac{1}{5}$ OR 0.2   **[1]**
5. $a = 13$ **[1]**; $b = 14$ **[1]**

1. a) $\frac{1}{2} \times 6 \times 8 \times 9$ **[1]**; 216cm² **[1]**

   b) $\sqrt[3]{216}$ **[1]**; 6×12 **[1]**; = 72cm **[1]**

2. $\pi r^2 = 2\pi r$ **[1]**; $r = 2$ **[1]**

3. $\frac{1}{2}(2x + x) \times 3x \times 20 = 900$ **[1]**;

   $9x^2 = 90$ **[1]**; $x^2 = 10$ **[1]**;

   $x = \sqrt{10}$ or 3.16cm **[1]**

4. $75 = 4 \times \pi \times r^2$ **[1]**;

   $r^2 = \frac{75}{4\pi}$ **[1]**; $r = 2.44$cm (to 3 significant figures) **[1]**

Pages 88–93 Practice Questions

Page 88–89

1. $m = 3$ **[1]**; $5 = 3 + c$, $c = 2$ **[1]**; $y = 3x + 2$ **[1]**
2. Table correctly complete (deduct 1 mark for one error and 2 marks for two errors; 0 marks for more than two errors) **[3]**

Gradient of Line	Gradient of Parallel Line	Gradient of Perpendicular Line
5	5	$-\frac{1}{5}$
3	3	$-\frac{1}{3}$
4	4	$-\frac{1}{4}$
$\frac{8}{9}$	$\frac{8}{9}$	$-\frac{9}{8}$

3. a) Esmai **[1]**
   b) 6mph **[2]** (1 mark for 2 miles in 20 minutes)
   c) 15–20 minutes **[1]**; steepest part of the line **[1]**
   d) 12.5 minutes **[1]**
   e) Esmai and Naval are level for first 5 minutes **[1]**; Esmai then speeds up and overtakes Naval **[1]**; Esmai wins, Naval is second and Gemma is third. **[1]**
4. a) Sketch showing the two curved lines **[1]** intersecting the y-axis at (0, 1) **[1]**

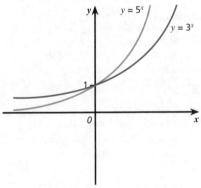

   b) Both pass through point (0, 1) **[1]**

5. a) 16 ÷ 10 **[1]**; = 1.6 **[1]**
   b) £1 = $1.6 **[1]**
6. a) Gradient of radius = $-\frac{5}{4}$ **[1]**;

   Gradient of tangent = $\frac{4}{5}$ **[1]**;

   $-5 = \frac{4}{5} \times 4 + c$, $c = -\frac{41}{5}$ **[1]**;

   $y = \frac{4}{5}x - \frac{41}{5}$ OR $5y = 4x - 41$ **[1]**

   b) $y = \frac{4}{5}x$ + any value **[1]**

7. Area of two triangles, plus a rectangle

   $= \frac{1}{24} + \frac{5}{12} + \frac{5}{24} = \frac{16}{24}$ mile **[1]**;

   Area of trapezium = $\frac{1}{3} = \frac{8}{24}$ mile **[1]**;

   distance travelled = 1 mile **[1]**

   Remember to change the unit of time to hours.

Page 90

1. $-4x > 20$ **[1]**; $x < -5$ **[1]**

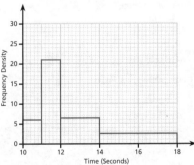

   $x < -5$

   If x is negative the inequality sign will change direction.

2. $x + y \geqslant 5$ **[1]**
3. $0 \leqslant p \leqslant 52$ where p is the number of passengers **[2]**

1. Angle ACB = Angle BDC = 90° **[1]**; Angle ABC = Angle DBC (the angle is common to both triangles) **[1]**; Angle BAC = Angle BCD (180° – Angle B – 90°), so the triangles are similar (three matching angles) **[1]**
2. 64 : 1 **[1]**

Page 91

1. $7.6^2 + 6.7^2 = y^2$ **[1]**; $57.76 + 44.89 = y^2$ **[1]**; $y = \sqrt{(102.65)} = 10.13$km **[1]**
2. $4^2 - 2^2 = f^2$ **[1]**; $16 - 4 = f^2$ **[1]**; $\sqrt{12} = f$, fence height = 3.46m **[1]**; No, because the fence height is 3.46m and Fang can only jump 3m. **[1]**
3. a) AC is 6cm (Pythagorean triple) **[1]**; $6^2 + 6^2 = AD^2$ **[1]**; $\sqrt{72} = AD = 8.49$cm (to 3 significant figures) **[1]**
   b) Scalene triangle **[1]**
4. $10^2 + 6^2 = b^2$ **[1]**; $100 + 36 = b^2$ **[1]**; $b = \sqrt{136} = 11.7$m (to 3 significant figures) **[1]**
5. $1.5^2 + 2^2 = 6.25$ **[1]**; $\sqrt{6.25} = 2.5$ **[1]**; It is a right-angled triangle because Pythagoras' Theorem applies **[1]**
6. $3^2 + 3^2 = d^2$ **[1]**; $d = \sqrt{18} = 4.24$cm **[1]**
7. $\sin 30° = \frac{x}{6}$ (where x is half the length of the hexagonal side) **[1]**; $x = 6 \times \sin 30° = 3$cm, so length of side = 6cm **[1]**
8. $\tan \theta = \frac{5}{3} = 1.6667$ **[1]**; $\theta = 59°$ **[1]**

9. a) Distance (x) of P from base of tower:

   $\tan 35° = \frac{14}{x}$, $x = \frac{14}{\tan 35°}$ **[1]**;

   $x = 19.9941$m **[1]**

   b) Distance (y) of Q from base of tower:

   $\tan 18° = \frac{14}{y}$ **[1]**; $y = 43.0876$m **[1]**

   c) $19.9941^2 + 43.0876^2 = PQ^2$, $399.764 + 1856.541 = PQ^2$ **[1]**; $PQ = 47.5$m **[1]**

Page 92

1. Area = $\frac{1}{2}bc \sin A$,

   $20 = \frac{1}{2} \times (5.7 \times 7.5 \times \sin A)$ **[1]**; $\sin A = \frac{40}{42.75} = 0.9357$ **[1]**; Angle $CAB = 69.34°$ **[1]**; Angle $CAB = 69°$ (to the nearest degree) **[1]**

2. $\frac{14}{\sin 50°} = \frac{12}{\sin A}$ **[1]**;

   $\sin A = \left(\frac{12 \times \sin 50°}{14}\right)$ **[1]**;

   Angle $BAC = 41.04°$ **[1]**

3. $a^2 = b^2 + c^2 - 2bc \cos A$ **[1]**; $a^2 = 80^2 + 30^2 - 2 \times 80 \times 30 \times \cos 30°$ **[1]**; $a^2 = 7300 - 4156.9$ **[1]**; $a = \sqrt{3143.1} = 56.06$m **[1]**; height of the Leaning Tower = 56m **[1]**

4. $\cos A = \frac{(b^2 + c^2 - a^2)}{2bc}$

   $= \frac{(220^2 + 250^2 - 78^2)}{2 \times 220 \times 250}$

   $= \frac{(104\,816)}{110\,000} = 0.95287...$ **[1]**;

   $A = 17.66°$ **[1]**

Page 93

1. a) 50 **[1]**
   b) Correctly drawn histogram, based upon frequency densities of 6, 21, 6.5 and 2.5 to give correct heights **[1]**; correct widths **[1]**; and correct labelling of axes. **[1]**

   c) 21 + 13 + 5 **[1]**; = 39 **[1]**
2. Median = 42 – 44 **[1]**; Q1 = 32 – 34 and Q3 = 52 – 54 **[1]**; correct scale **[1]**; and accurately drawn boxes **[1]**

Answers

3. a) Drawn and labelled axes **[1]**; and accurately plotted points. **[1]**

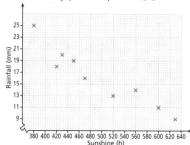

b) Negative correlation – as sunshine increases, rain decreases **[1]**

Pages 94–105 Revise Questions

Page 95 Quick Test
1. $1^2 \times 400 = 400$ OR $(0.9)^2 \times 400 = 324$
2. Width: $2.805 \leqslant w < 2.815$cm
   Length: $3.865 \leqslant l < 3.875$cm

Page 97 Quick Test
1. $x = -1.4$ or 3.4
2. $x = 1.85$ or -0.18
3. $x = 2 \pm \sqrt{2}$
4. $x = 5.19$ or -0.19

Page 99 Quick Test
1. $x = 2$, $y = 1$
2. $x = 4$, $y = 11$ and $x = -1$, $y = -4$
3. $x = 2$, $y = 2$ and $x = -1$, $y = -1$
4. $18x^2 + 12x + 7$

Page 101 Quick Test
1. $n + (n + 1) = 2n + 1$, $2n$ is even, therefore $2n + 1$ is odd.
2. n^2 and $(n + 1)^2$, n^2 and $n^2 + 2n + 1$. The difference is $2n + 1$, which is an odd number.
3. Set of integers is $n - 2$, $n - 1$, n, $n + 1$, $n + 2$. Mean $= \dfrac{5n}{5} = n$
4. $\dfrac{2}{x} + \dfrac{3}{x^2} = \dfrac{2x}{x^2} + \dfrac{3}{x^2} = \dfrac{2x + 3}{x^2}$

Page 103 Quick Test
1. **a)** $32°$
   b) $56°$
   c) $88°$
   d) $65°$

Page 105 Quick Test
1. **a)** N

b) N

c) N

d) N

e) N

Pages 106–111 Review Questions

Page 106–107
1. $\dfrac{1}{2} \times 2 \times 30 + \dfrac{1}{2} \times 1 \times 60$ **[1]**;

   $+ \dfrac{1}{2}(30 + 60) \times 1$ **[1]**; $30 + 30 + 45$

   $= 105$ miles **[1]**

> Break the area under the curve down into two triangles and a trapezium.

2. $m_1 = \dfrac{3}{5}$, $m_2 = -\dfrac{5}{3}$ **[1]**; $\dfrac{3}{5} \times -\dfrac{5}{3} = -1$ **[1]**
3. **a) i)** A straight horizontal line **[1]**
   ii) 30 minutes **[1]**
   b) 4.30pm **[1]**
   c) i) $\dfrac{18}{0.5}$ **[1]**; 36km/h **[1]**
   ii) 6pm **[1]**
4. 4 (approx.) **[1]**
5. **a)** $r^2 = 3^2 + 5^2$ **[1]**; $r^2 = 34$ **[1]**;
   $x^2 + y^2 = 34$ **[1]**
   b) Gradient of radius $= \dfrac{5}{3}$ **[1]**; Gradient of tangent $= -\dfrac{3}{5}$ **[1]**; $5y + 3x = 34$ or equivalent **[1]**
6. **a)**

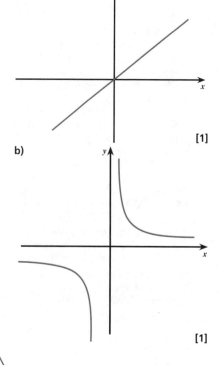

[1]

b)

[1]

7. **a)** $(x - 3)(x + 3) = x^2 - 9$ **[1]**;
   $(x^2 - 9)(x - 2)$ **[1]**; $x^3 - 2x^2 - 9x + 18$ **[1]**
   b)

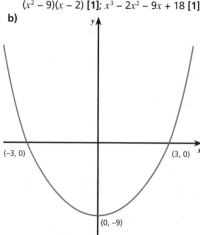

Correct shape graph **[1]**; correct points of intersection **[1]**

8. $m_1 = -\dfrac{1}{2}$, $m_2 = 2$ **[1]**; $6 = 2 \times 3 + c$, $c = 0$ **[1]**; $y = 2x$ **[1]**

Page 108
1. $n = 3, 4, 5, 6, 7$ **[1]**
2. $2x > 8$ **[1]**; $x > 4$ **[1]**
3. $4 \leqslant y \leqslant 12$ **[1]**; 4, 5, 6, 7, 8, 9, 10, 11, 12 **[1]**
4. $(x + 8)(x - 3) \geqslant 0$ **[1]**; one solution is $x \leqslant -8$ **[1]**; one solution is $x \geqslant 3$ **[1]**
5. $-9x > 1$, $x < -\dfrac{1}{9}$ **[1]**

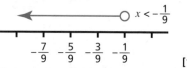

[1]

6. If t is number of televisions, $t > 6$ and $t \leqslant 20$ **[1]**; $6 < t \leqslant 20$ **[1]**

1. $56°$ **[1]**; $64°$ **[1]**; $60°$ **[1]**
2. **a)** Isosceles **[1]**
   b) Prove using SSS OR SAS **[3]**
   c) $AD^2 = 10^2 - 6^2$ **[1]**; $AD = 8$cm **[1]**;
   Area $= \dfrac{1}{2}(12 \times 8) = 48$cm² **[1]**

Page 109
1. Diagonal length of base:
   $200^2 + 200^2 = d^2$ **[1]**; $d = \sqrt{80000} = 282.84$m; $BC = 141.42$m **[1]**;
   Using triangle TCB: $310^2 - 141.42^2 = TC^2$ **[1]**; $TC = \sqrt{76\,100.38}$, $TC = 276$m **[1]**
2. 5cm **[1]**
3. Third side of the triangle = 5cm (Pythagorean Triple) **[1]**;
   Area $= \dfrac{1}{2}$ base × height, Area $= 6 \times 5$ **[1]**;
   Area = 30cm² **[1]**
4. **a)** $8^2 + 5^2 = CE^2$ **[1]**; $CE = 9.434$cm **[1]**;
   $\cos \theta = \dfrac{4.717}{13} = 0.36285$ **[1]**; $\theta = $ Angle $PCE = 68.7°$ **[1]**
   b) $13^2 - 4.717^2 = h^2$ **[1]**; Height of P above base = 12.1cm **[1]**
5. $\sin 78° = \dfrac{x}{25}$ **[1]**; $x = 25 \times \sin 78°$
   $= 24.45$m **[1]**

Answers

Page 110

1. a) Angle C is the smallest [1], because it is opposite the shortest side. [1]

   b) $\cos C = \dfrac{(a^2 + b^2 - c^2)}{2ab}$ [1];

   $\cos C = \dfrac{(13^2 + 11^2 - 10^2)}{2 \times 13 \times 11}$ [1]; $\cos C$

   $= 0.6643$ [1]; Angle $C = 48°$ (to the nearest degree) [1]

   c) Area $= \frac{1}{2}ab \sin C$ [1];

   Area $= \frac{1}{2} \times 13 \times 11 \times \sin 48°$ [1];

   Area $= 53.13\text{cm}^2$ [1]

2. $b^2 = a^2 + c^2 - 2ac \cos B$ [1];

   $b^2 = 144 + 196 - 2 \times 12 \times 14 \times \cos 50°$ [1];

   $= 340 - 215.976\,636\,9$

   $= 124.023\,363\,1$ [1]; $b = 11.1\text{km}$ [1] OR

   $\dfrac{14}{\sin 75°} = \dfrac{b}{\sin 50°}$ [1]; $b = \dfrac{(14 \times \sin 50°)}{\sin 75°}$ [1];

   $= \dfrac{10.724\,6222}{0.965\,925\,826}$ [1]; $= 11.1\text{km}$ [1]

Page 111

1. a) All frequency values in table correct. Calculated using Frequency Density × Class Width [2]

Price (pounds)	Frequency
$0 < P \leqslant 5$	50
$5 < P \leqslant 10$	90
$10 < P \leqslant 20$	70
$20 < P \leqslant 40$	60

   b) $2.5 \times 50 + 7.5 \times 90 + 15 \times 70 + 30 \times 60$

   [1]; $\dfrac{3650}{270}$ [1]; £13.52 [1]

   c) $\dfrac{270 + 1}{2} = 135.5$ [1]; $5 < P \leqslant 10$ [1]

2. a) It is biased because she only speaks to library users. [1]

   b) A random [1] selection of 100 people who live in Malmesbury [1]

3. a) Line of best fit drawn as per diagram below [1]; Accept 120 to 135 [1]

   b) The estimate is not reliable as it is outside the data range. [1]

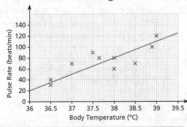

Pages 112–115 Practice Questions

Page 112

1. $\dfrac{500 \times 8}{0.2}$ [1]; $= 20\,000$ [1]

2. 5kg of carrots cost £4.45, so 1kg of carrots = 89p [1]; £3.56 + 5kg of potatoes = £6.11, so 1kg of potatoes = 51p [1]; 4kg of potatoes = £2.04 [1]

---

3. £657 $= (h \times 35) - 218$ [1]; $\dfrac{657 + 218}{35}$ [1];

   $= 25$ hours [1]

4. $\sqrt{(6.2^2 - 3.6)} = 5.902\,541\,825$,

   $2.6 \times 0.15 = 0.39$ [1];

   $5.902\,541\,825 \div 0.39 = 15.134\,722\,63$ [1]

   a) 15.13 [1]

   b) 15.1 [1]

5. $\dfrac{£560}{117.5} \times 100$ [1]; $= £476.60$ [1]

6. a) $5093 + 3x = 5315$ [1];

   $\dfrac{5315 - 5093}{3} = \dfrac{222\text{ pence}}{3} = 74$ units [1]

   b) $\dfrac{3}{100} \times £50.93 = £1.5279$ [1];

   $£50.93 + £1.5279 = £52.46$ [1]

Page 113

1. $x = \pm 3$ [1]

2. a) $(x + 4)^2 - 28$, $p = 4$ [1]; $q = -28$ [1]

   b) $x = -4 \pm \sqrt{28}$ [1]; $x = -4 \pm 2\sqrt{7}$ [1]

   c) $(-4, -28)$ [1]

3. a) $A = \frac{1}{2}(a + b)h$,

   $14 = \frac{1}{2}(4x - 6 + 2x + 1)\,2x$ [1];

   $14 = \frac{1}{2}(12x^2 - 10x)$ [1];

   $6x^2 - 5x - 14 = 0$ [1]

   b) $(6x + 7)(x - 2)$ [1]; $x = 2$ [1]

   If x is a length it must be positive.

1. a) Straight-line graph correct [1]; and curved graph correct [1]

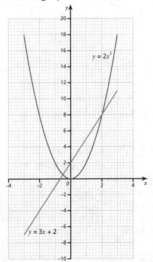

   b) $(-0.5, 0.5)$ [1]; $(2, 8)$ [1]

   When graphs of simultaneous equations are plotted, the x-coordinates of the points of intersection give you your solutions.

2. $y = 1 - 2x$ [1]; $1 - 2x = x^2 - 2$ [1];

   $x^2 + 2x - 3 = 0$, $(x + 3)(x - 1) = 0$ [1];

   $(-3, 7)$, $(1, -1)$ OR $x = -3$, $y = 7$ or $x = 1$, $y = -1$ [1]

---

3. a) $f(3x - 2) = (3x - 2)^2 + 1$ [1];

   $fg(x) = 9x^2 - 12x + 5$ [1]

   b) $gf(x) = 3(x^2 + 1) - 2$ [1];

   $= 3x^2 + 1 \neq 9x^2 + 12x + 5$ [1]

   c) $9x^2 - 12x - 5 = 0$ [1];

   $x = \dfrac{12 \pm \sqrt{(-12)^2 - 4 \times 9 \times (-5)}}{18}$ [1];

   $x = \dfrac{5}{3}$ or $-\dfrac{1}{3}$ [1]

Page 114

1. $n^2 + (n + 1)^2 + (n + 2)^2$ [1]; $3n^2 + 6n + 5$ [1];

   $3(n^2 + 2n + 2) - 1$, which is 1 less than a multiple of 3 [1]

2. Let $n =$ chosen number [1];

   $\dfrac{3n + 30}{3} = n + 10$ [1]; $n + 10 - n = 10$ [1];

3. $\dfrac{3}{x + 2} + \dfrac{9}{(x + 2)(x - 1)}$ [1]; $\dfrac{3(x - 1) + 9}{(x + 2)(x - 1)}$ [1];

   $\dfrac{3x - 3 + 9}{(x + 2)(x - 1)} = \dfrac{3(x + 2)}{(x + 2)(x - 1)} = \dfrac{3}{(x - 1)}$ [1]

4. $\dfrac{6(3x - 2)}{(3x + 2)(3x - 2)} + \dfrac{4(3x + 2)}{(3x + 2)(3x - 2)}$ [1];

   $\dfrac{18x - 12 + 12x + 8}{9x^2 - 4}$ [1]; $\dfrac{30x - 4}{9x^2 - 4}$ [1]

   Find a common denominator.

5. Area of square $= 36x^2$ [1];

   Area of circle $= \pi(3x)^2 = 9\pi x^2$ [1];

   $36x^2 : 9\pi x^2$ [1]; $4 : \pi$ [1]

6. $(16n^2 + 8n + 1) - (16n^2 - 8n + 1)$ [1];

   $16n$ [1]; $8(2n)$ which is a multiple of 8 [1]

7. $(2n + 2)^2 - (2n)^2$ [1]; $8n + 4$ [1]; $4(2n + 1)$ which is a multiple of 4 [1]

8. $a + b = ab$ [1]; $b + 2 + b = b(b + 2)$ [1];

   $b^2 = 2$, therefore b is not an integer, and $a = \sqrt{2} + 2$, so a is not an integer. [1]

Page 115

1. a) $a = 34°$ $(90° - 56°)$ [1]

   b) $b = 102°$ $(204° \div 2)$ [1]; $c = 78°$ $(180° - 102°)$ [1]

   c) $d = 72°$ (alternate segment) [1]; $e = 50°$ $(180° - 72° - 58°)$ [1]

   d) $f = 37°$ (angle subtended by a chord) [1]; $g = 21°$ (angle subtended by a chord) [1]

   e) $h = 42°$ $(90° - ((180° - 84°) \div 2)$ $= 90° - 48° = 42°$ [1]; $i = 42°$ (isosceles triangle) [1]

   f) $j = 36°$ (isosceles triangle so $5j = 180°$) [1]

1. a) $a + \frac{1}{2}(c - a)$ [2];

   $\frac{1}{2}a + \frac{1}{2}c = \frac{1}{2}(a + c)$ [1]

   b) $\overrightarrow{DE} = \frac{1}{2}a + \frac{1}{2}(c - a)$ [1];

   $= \frac{1}{2}a + \frac{1}{2}c - \frac{1}{2}a = \frac{1}{2}c$ [1];

   $\overrightarrow{AC} = c$, so $\overrightarrow{DE}$ is parallel to $\overrightarrow{AC}$ [1]

Answers

2. a) Scalar **[1]**
b) Vector **[1]**
c) Vector **[1]**

Pages 116–119 Review Questions

Page 116

1. $805.975 \div 6.14 = 131.266\,286\,6$ **[1]**
a) 131.27 **[1]**
b) 131 **[1]**

2. a) Per day £84 (adults) + £35.70 (child)
= £119.70 **[1]**; Holiday costs
$9 \times £119.70$ **[1]**; = £1077.30 **[1]**
b) 5% of £1077.30 **[1]**; = £53.87 **[1]**
c) 11.10am + 4 hours 20 minutes + 1
hour 34 minutes, flight arrives in
Turkey at 5.04pm (17:04) in English
time **[1]**; Turkish arrival time =
7.04pm (19:04) **[1]**

3. 18.0625cm^2 **[1]**

4. a) Option 1: £146 + £80 × 12 = £1106,
$\frac{95}{100} \times £1106 = £1050.70$ **[1]**;
Option 2: £548 + £43 × 12 = £1064
$\frac{95}{100} \times £1064 = £1010.80$ **[1]**;
Option 2 is cheaper **[1]**
b) £1050.70 – £1010.80 = £39.90 **[1]**

Page 117

1. a) $y = 6$ **[1]**
b) $x^2 + (3)^2 = 36$, $x^2 = 27$ **[1]**; $x = 5.2$ or
-5.2 **[1]**; $(-5.2, 3)$ and $(5.2, 3)$ **[1]**

2. a) $(x - 6)^2 - 10$ **[1]**; $p = -6$ **[1]**; $q = -10$ **[1]**;
b) $(x - 6)^2 = 10$, $x - 6 = \pm\sqrt{10}$ **[1]**;
$x = 6 + \sqrt{10}$, $x = 6 - \sqrt{10}$ **[1]**

3. $\frac{4(x + 2)}{x(x + 2)} + \frac{4x}{x(x + 2)} = 3$ **[1]**; $\frac{8x + 8}{x^2 + 2x} = 3$,
$8x + 8 = 3x^2 + 6x$ **[1]**;
$3x^2 - 2x - 8 = 0$ **[1]**; $(3x + 4)(x - 2)$,
$x = -\frac{4}{3}$ and $x = 2$ **[1]**

4. $\frac{1}{2} \times \frac{x}{5} \times \frac{x}{15} = 4$ **[1]**; $x^2 = 600$ **[1]**;
$x = \sqrt{600}$, $x = 10\sqrt{6}$ **[1]**

5. $x = \frac{7 \pm \sqrt{(-7)^2 - 4 \times 1 \times -15}}{2}$ **[1]**;
$x = -1.72$ and 8.72 (to 3 significant
figures) **[1]**

1. a) Correct curve **[1]**; correct straight
line **[1]**

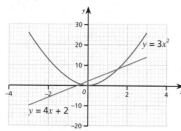

b) $x = 1.6$ to 1.8 **[1]**; (-0.3) to (-0.5) **[1]**

2. $y = x + 4$, $x + 4 = x^2 + 2$ **[1]**;
$x^2 - x - 2 = 0$ **[1]**; $(x - 2)(x + 1) = 0$,
$x = 2$, $y = 6$ **[1]**;
$x = -1$, $y = 3$ **[1]**

Page 118

1. a) $3n + 1$ **[1]**
b) $(3n + 1)(3(n + 1) + 1)$ **[1]**;
$9n^2 + 15n + 4$ **[1]**;
$3(3n^2 + 5n + 1) + 1$ **[1]**

2. $(n + 1)^2 - n^2$ **[1]**; $= n^2 + 2n + 1 - n^2$
$= 2n + 1$ **[1]**; $= n + (n + 1)$ **[1]**

3. $\frac{(x + 3)(x + 2)}{x^2 + 2x}$ **[1]**; $\frac{(x + 3)(x + 2)}{x(x + 2)}$ **[1]**;
$\frac{x + 3}{x}$ **[1]**

4. Any odd number –7 **[1]**

5. Area = x^2, diagonal = $\sqrt{x^2 + x^2}$
$= \sqrt{2x^2}$ **[1]**; $\sqrt{2x^2} \times \sqrt{2x^2} = 2x^2$,
$\frac{2x^2}{2} = x^2$ **[1]**

6. $9n^2 + 6n + 1 - (9n^2 - 6n + 1)$ **[1]**;
$12n$ **[1]**; $4(3n)$ **[1]**

7. a) $n^2 + 1$ **[1]**
b) $(n^2 + 1) + ((n + 1)^2 + 1)$ **[1]**;
$2n^2 + 2n + 3$ **[1]**; $2(n^2 + n + 1) + 1$ **[1]**

8. $\frac{2}{x - 2} - \frac{8}{(x + 2)(x - 2)}$ **[1]**; $\frac{2(x + 2)}{(x - 2)(x + 2)}$
$- \frac{8}{(x + 2)(x - 2)}$, $\frac{2x - 4}{(x + 2)(x - 2)}$ **[1]**;
$\frac{2(x - 2)}{(x + 2)(x - 2)}$, $\frac{2}{(x + 2)}$ **[1]**

9. $(2n + 1)^2$ **[1]** $= 4n^2 + 4n + 1$ **[1]**;
$4(n^2 + n) + 1$ **[1]**

10. The sum of any two primes that are
not 2, e.g. 3 + 5 = 8 **[1]**

Page 119

1. a) $k = 52°$ (alternate segment) **[1]**
b) $l = 90°$ (angle subtended from
diameter) **[1]**
c) $m = 48°$ (96° ÷ 2) **[1]**; $n = 42°$
((180° – 96°) ÷ 2) **[1]**

2. a) a = radius **[1]**; b = circumference **[1]**;
c = arc **[1]**; d = segment **[1]**;
e = sector **[1]**
b) Arc length = $\frac{\theta}{360} \times 2 \times \pi \times r$ **[1]**;
Arc length = $\frac{40}{360} \times 2 \times 3.14 \times 10$
= 6.977 777 778 **[1]**;
Arc length = 6.98cm **[1]**

1. a) $\overrightarrow{AC} = 4\mathbf{c} - 4\mathbf{a}$ **[1]**
b) $\overrightarrow{AP} = \frac{1}{4}(4\mathbf{c} - 4\mathbf{a}) = \mathbf{c} - \mathbf{a}$ **[1]**
c) $\overrightarrow{DP} = 4\mathbf{a} + \mathbf{c} - \mathbf{a} = 3\mathbf{a} + \mathbf{c}$ **[1]**

2. a) True **[1]**
b) False **[1]**

Pages 120–131 Mixed Exam-Style Questions

1. Trapezium below line C shaded **[1]**
2. $3x^2(x + 2)$ **[1]**
3. $3x + 9 - 5 = 2x - 4$ **[1]**; $3x + 4$
$= 2x - 4$ **[1]**; $x = -8$ **[1]**
4. $\frac{5}{7}\left(\frac{1}{5} - \frac{2}{5}\right) + \frac{18}{7}$ **[1]**; $\frac{17}{7}$ **[1]**
5. $-5x^2 + 10x - 2y$ **[2]**
6. $4p^2y + 6py$ **[2]**
7. $x - 3 = -\frac{3}{5}$ **[1]**; $x = \frac{12}{5}$ **[1]**
8. $a = 6$, $b = -16$, $c = -14$, $d = 24$,
$6x^3 - 16x - 14x + 24$ **[2]** (1 mark for any
two correct terms)
9. a) $(x - 2)(x + 1)$ **[1]**
b) $(2x - 1)(x - 2)$ **[1]**
c) $\frac{(2x - 1)(x - 2)}{(x - 2)(x + 1)}$ **[1]**; $\frac{2x - 1}{x + 1}$ **[1]**
10. $30 = 60 - q^2$ **[1]**; $q = \sqrt{30}$ **[1]**

> The question asks for an exact value,
> so leave your answer in surd form.

11. a) $(x + 4)(x - 4)$ **[1]**
b) $x = 4$ **[1]**; $x = -4$ **[1]**
12. Ethan is correct **[1]**; $r^2 = \frac{25}{\pi}$,
$r = \sqrt{\frac{25}{\pi}} = \frac{5}{\sqrt{\pi}}$ **[1]**

> You must square root both the
> numerator and denominator.

13. $\left(\sqrt{5}\right)^2 + 3\sqrt{5}$ **[1]**; $5 + 3\sqrt{5}$ **[1]**
14. 6.677, 6.767, 6.77, 6.8 **[1]**
15. a) $5 \times 3 \times 3$ **[1]**; $= 5 \times 3^2$ **[1]**
b) $3 \times 5 \times 7$ **[1]**
c) 3×5 **[1]**; = 15 **[1]**
d) x^2y^2 **[2]**
16. $\frac{31}{6} - \frac{7}{3}$ **[1]**; $\frac{31}{6} - \frac{14}{6}$ **[1]**; $\frac{17}{6}$ **[1]**
17. $P = 1.1 \times 1.1$ **[1]** = 1.21, 21% increase **[1]**
18. Dave's Dongles: £16.20 for six months **[1]**;
£18 for following six months, total
£205.20 **[1]**; Ian's Internet: £14.62 for
four months **[1]**; £17.20 for following
8 months, total £196.08 **[1]**; Ian's
Internet is cheaper **[1]**
19. a) 1408.44 **[1]**
b) 0.140844 **[1]**
20. $1.5 \times 10^8 \div 3.5 \times 10^4$ **[1]**;
4285.7 (to 1 d.p.) = 4.2857×10^3 **[1]**
21. Speed = $\frac{\text{Distance}}{\text{Time}}$ **[1]**
22. Exterior angle = $\frac{360°}{20°}$ = 18° **[1]**;
interior angle = 180° – 18° = 162° **[1]**
23. $3n + 5$ **[1]**

24. a) $2 \leqslant a < 4$

b) $(1 \times 256) + (3 \times 219) + (5 \times 165) +$
$(7 \times 120) + (9 \times 65) = 3163$ **[1]**;
$3163 \div 825$ **[1]**; $= 3.83$ (to 2 d.p.) **[1]**

> Remember to use midpoints to
> calculate the mean of grouped data.

c)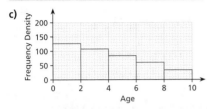

Correct widths **[1]**; correct frequency
densities: 128, 109.5, 82.5, 60, 32.5 **[1]**

d) No **[1]**; the median is now the 452nd
value, which is in the same group. **[1]**

25. a) $F \alpha \dfrac{k}{x^2}$, $4 = \dfrac{k}{9}$ **[1]**; $k = 36$,

$F = \dfrac{36}{x^2}$ **[1]**

b) $F = \dfrac{36}{25}$ or 1.44 **[1]**

c) $20 = \dfrac{36}{x^2}$ **[1]**; $x^2 = \dfrac{36}{20}$,

$x = \sqrt{\dfrac{9}{5}}$ or 1.34 **[1]**

26. a) $x^2 = 6^2 + 9^2 - 2 \times 6 \times 9 \times \cos 37°$ **[1]**;
$x^2 = 30.74\ldots$**[1]**; $x = 5.55$cm
(to 3 significant figures) **[1]**

b) $\dfrac{1}{2} \times 6 \times 9 \times \sin 37°$ **[1]**; $= 16.25$cm² **[1]**

c) 9 **[1]**

27. $x = 0$ **[1]**

28. a) $3 : 8$ **[1]**

b) $\dfrac{20}{74} \times 100$ **[1]**; $= 27\%$
(to the nearest 1%) **[1]**

29. a)

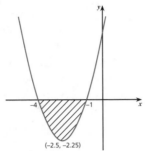

[1]

b) $(x + 1)(x + 4) \leqslant 0$ **[1]**;
$-4 \leqslant x \leqslant -1$ **[1]**

c) Shaded region on graph above **[1]**

30. a) i) $2b - 2a$ **[1]**

ii) $2a$ **[2]**

b) $4a - b$ **[2]**

c) $\overrightarrow{DY} = 4a - \dfrac{2}{3}(2b - 2a)$
$= \dfrac{16}{3}a - \dfrac{4}{3}b$ **[1]**;
$\overrightarrow{DX} = 4a - b$ **[1]**;
$DY = \dfrac{4}{3}DX$, therefore,
DXY is a straight line **[1]**

31. $x^2 - 1 = 3x + 3$ **[1]**;
$(x - 4)(x + 1) = 0$ **[1]**;
$x = 4, y = 15$ **[1]**;
$x = -1, y = 0$ **[1]**

32. a) 33° **[1]**; radius and tangent meet at
90° **[1]**

b) 57° **[1]**; Triangle in semicircle, angle
at circumference = 90° or alternate
segment theorem **[1]**

33. $\dfrac{1}{5}t^2$ **[2]**

34. $2 \times \pi \times 6 = 12\pi$ **[1]**; $12\pi \times \dfrac{115}{360}$ **[1]**;
$12\pi \times \dfrac{115}{360} + 12$ **[1]**; 24.0 (to 3 significant
figures) **[1]**

35. $\dfrac{6}{11} \times \dfrac{5}{11}$ **[1]**; $\dfrac{30}{121} \times 2$ **[1]**; $\dfrac{60}{121}$ **[1]**

> The bead is replaced each time.

36. a) 1.55×2.655 **[1]**; $= 4.11525$ **[1]**

b) $1.45 \div 2.655$ **[1]**; $= 0.546$ (to 3 d.p.) **[1]**

37. a) $(-2, 0)$ **[1]**; $(-1, 0)$ **[1]**

b) $(x + 1.5)^2 - 0.25$ **[2]**; $(-1.5, -0.25)$ **[1]**

c) $(-3.5, -0.25)$ **[1]**

Glossary and Index

Collins

Maths

Edexcel GCSE Revision

Maths

Higher

Higher

Edexcel GCSE

Workbook

**Linda Couchman
and Rebecca Evans**

Revision Tips

Rethink Revision

Have you ever taken part in a quiz and thought *'I know this!'* but, despite frantically racking your brain, you just couldn't come up with the answer?

It's very frustrating when this happens but, in a fun situation, it doesn't really matter. However, in your GCSE exams, it will be essential that you can recall the relevant information quickly when you need to.

Most students think that revision is about making sure you *know* stuff. Of course, this is important, but it is also about becoming confident that you can **retain** that *stuff* over time and **recall** it quickly when needed.

Revision That Really Works

Experts have discovered that there are two techniques that help with all of these things and consistently produce better results in exams compared to other revision techniques.

Applying these techniques to your GCSE revision will ensure you get better results in your exams and will have all the relevant knowledge at your fingertips when you start studying for further qualifications, like AS and A Levels, or begin work.

It really isn't rocket science either – you simply need to:

- **test yourself** on each topic as many times as possible
- **leave a gap** between the test sessions.

Three Essential Revision Tips

1. **Use Your Time Wisely**

   - Allow yourself plenty of time.
   - Try to start revising at least six months before your exams – it's more effective and less stressful.
   - Your revision time is precious so use it wisely – using the techniques described on this page will ensure you revise effectively and efficiently and get the best results.
   - Don't waste time re-reading the same information over and over again – it's time-consuming and not effective!

2. **Make a Plan**

   - Identify all the topics you need to revise (this Complete Revision & Practice book will help you).
   - Plan at least five sessions for each topic.
   - One hour should be ample time to test yourself on the key ideas for a topic.
   - Spread out the practice sessions for each topic – the optimum time to leave between each session is about one month but, if this isn't possible, just make the gaps as big as realistically possible.

3. **Test Yourself**

   - Methods for testing yourself include: quizzes, practice questions, flashcards, past papers, explaining a topic to someone else, etc.
   - This Complete Revision & Practice book provides seven practice opportunities per topic.
   - Don't worry if you get an answer wrong – provided you check what the correct answer is, you are more likely to get the same or similar questions right in future!

Visit our website to download your free flashcards, for more information about the benefits of these techniques, and for further guidance on how to plan ahead and make them work for you.

www.collins.co.uk/collinsGCSErevision

Contents

 Number   Algebra 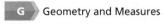 Geometry and Measures

 Statistics   Probability R Ratio, Proportion and Rates of Change

Order and Value

1 **a)** Write two hundred million in standard form.

Answer _____ [1]

b) Write 6.78×10^{-4} as an ordinary number.

Answer _____ [1]

2 Work out $(1.5 \times 10^4) + (3.5 \times 10^3)$. Give your answer in standard form.

Answer _____ [3]

3 Work out the value of $\dfrac{(3\sqrt{m})}{y}$ where $m = 8.1 \times 10^3$ and $y = 2.7 \times 10^{-2}$

Give your answer in standard form.

Answer _____ [3]

4 If $45 \times 82 = 3690$, work out the value of:

a) 4.5×8.2

Answer _____ [1]

b) 0.045×0.82

Answer _____ [1]

5 Work out $0.8645 \div 0.5$

Answer _____ [2]

6 If $53.\blacktriangle4 \times 0.2 = 10.668$, work out the value of $\blacktriangle$

Answer _____ [1]

Total Marks _____ / 13

Types of Number

1. 3, 7, 9, 12, 16, 20, 31

   From this list choose:

   a) Three prime numbers.   Answer _____ [1]

   b) Two numbers that are factors of 21.   Answer _____ [1]

   c) Three numbers that are multiples of 4.   Answer _____ [1]

   d) Two square numbers.   Answer _____ [1]

   e) The square root of 400.   Answer _____ [1]

2. Write 76 as a product of prime factors.

   Answer _____ [2]

3. Find the highest common factor (HCF) of 684 and 468.

   Answer _____ [3]

4. Subtract the sum of all the odd numbers from 1 to 999 from the sum of all the even numbers from 2 to 1000.

   Answer _____ [2]

5. 3797 is a special prime number because 379, 37 and 3 are all prime.

   Is 2797 a special prime number? Explain your answer.

   Answer _____ [2]

Basic Algebra

1 Solve the equation $\dfrac{2x + 4}{4} = 2$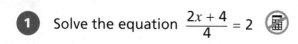

Answer _____ [3]

2 Work out the value of $4xy - x^2$ when $x = -2$ and $y = 7$.

Answer _____ [1]

3 Expand and simplify $3x(x - y) + y(x + 5)$

Answer _____ [2]

4 Solve $4x + 7 = 6x - 5$

Answer _____ [2]

5 Solve $\dfrac{2}{x} - 6 = 12$

Answer _____ [2]

6 Write $9xy - 3y^2 + 6x^2y$ in the form $ay(bx + cy + dx^2)$, where a, b, c and d are integers.

Answer _____ [2]

Total Marks _____ / 12

Factorisation and Formulae

1 Expand $(x + 4)(x - 2)$

Answer _____ [2]

2 Factorise $2x^2 + 5x + 2$

Answer _____ [2]

3 The formula below links velocity, time and acceleration:

$v = u + at$

a) Use the formula to find the value of v when $u = 15$, $a = 2.5$ and $t = 10$.

Answer _____ [1]

b) Rearrange to make t the subject of the formula.

Answer _____ [2]

c) Find the value of t when $v = 25$, $a = 1.6$ and $u = 11$.

Answer _____ [1]

4 Rearrange the formula to make r the subject:

$p = \dfrac{3r - 1}{r + 2}$

Answer _____ [3]

Total Marks _____ / 11

Ratio and Proportion

1 If $\frac{2}{7}$ of the pupils in a class are girls, what is the ratio of boys to girls?

Answer _____ [1]

2 A coach uses 11 litres of fuel to travel 161.7km. How far can it travel on 13 litres of fuel?

Answer _____ [2]

3 £700 is divided between Sarah, John and James. Sarah has twice as much as John and John has three times as much as James.

How much does Sarah receive?

Answer _____ [2]

4 Simplify 15 millilitres : 3 litres

Answer _____ [1]

5 When an apple falls from a tree, the distance (d) that it falls is proportional to the square of the time (t) taken to reach the ground.

If $d = 125$ metres when $t = 5$ seconds, work out:

a) The constant of proportionality.

Answer _____ [2]

b) The time taken for the apple to fall 48 metres. Give your answer to 1 decimal place.

Answer _____ [2]

Total Marks _____ / 10

Variation and Compound Measures

1 £4000 is invested at 1.6% compound interest per annum.

Work out its value after three years. Give your answer to the nearest pound.

Answer _____ [3]

2 A greyhound runs 77.8 metres in eight seconds. What is the greyhound's average speed in:

a) Metres per second (to 3 decimal places)?

Answer _____ [2]

b) Kilometres per hour (to 2 decimal places)?

Answer _____ [2]

3 A silver ring weighing 2g has a density of 10.49g/cm³.

Work out the volume of silver in the ring.
Give your answer to an appropriate degree of accuracy.

Answer _____ [2]

4 The force (F) between two magnets is inversely proportional to the square of the distance (d) between them.

Work out the constant of proportionality if $F = 12$ when $d = 3$.

Answer _____ [3]

Total Marks _____ / 12

Angles and Shapes 1 & 2

1 Work out the value of x.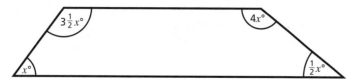

Answer _____ [2]

2 The interior angle of a regular polygon is 150°.

Work out how many sides the polygon has. 📱

Answer _____ [2]

3 A helicopter leaves its base and flies 40km on a bearing of 050° and then 30km on a bearing of 105°. 📱

a) Draw a scale diagram to show this information. How far is the helicopter from its base?

Answer _____ [2]

b) On what bearing does the helicopter need to fly in order to return to its base?

Answer _____ [1]

Total Marks _____ / 7

Fractions

1 Write down $\frac{45}{63}$ in its simplest form. 📱

Answer _____ [1]

2 Neesha ate $\frac{2}{3}$ of her chocolate bar this morning and then ate $\frac{3}{5}$ of what was left in the afternoon. How much is left to eat tomorrow? 📱

Answer _____ [2]

3 Which is larger, $\frac{7}{9}$ of 81 or $\frac{2}{7}$ of 217? You must show your working. 📱

Answer _____ [2]

4 Express 14 minutes as a fraction of 2.4 hours. Give your answer in the simplest form. 📱

Answer _____ [2]

5 Change $0.2\dot{7}$ to a fraction in its simplest form. 📱

Answer _____ [3]

6 Find the value of p if $\left(\frac{3}{2}\right)^{p} - \frac{3}{2} = \frac{3}{4}$ 📱

Answer _____ [2]

Total Marks _____ / 12

Percentages

1 In five months, a population of rats increases in number by 20% and then by 35%.

If there were 150 rats originally, how many are there at the end of the five-month period?

Answer _____ [2]

2 Write down 18g as a percentage of 0.075kg.

Answer _____ [2]

3 In a small village school, 22% of the children caught chicken pox.

If 11 children caught chicken pox, how many children attended the school?

Answer _____ [2]

4 This year, Pratik grew 31 tomato plants. This is a 38% reduction on last year.

How many tomato plants did Pratik grow last year?

Answer _____ [2]

5 Temi paid tax on £14 000 at 24%. She paid the tax in 12 equal monthly instalments.

Work out how much tax she paid each month.

Answer _____ [1]

Total Marks _____ / 9

Probability 1 & 2

1 A spinner has five sides: red, blue, yellow, green and pink.

The table below shows the probability associated with each colour.

Colour	Red	Blue	Yellow	Green	Pink
Probability	x	0.3	x	$3x$	0.2

a) Find the value of x.

Answer _____ [3]

b) Is the spinner fair? Give a reason for your answer.

_____ [2]

2 A bag contains n counters. Five of the counters are blue, three are yellow and the rest are red.

A counter is taken from the bag at random.

a) Write an expression in terms of n to represent the probability that the counter is red.

Answer _____ [2]

b) The counter is replaced and two counters are taken from the bag at random.

Show that the probability that one counter is blue and the other is yellow is $\dfrac{30}{n(n-1)}$

_____ [2]

3 A veterinary practice surveyed its clients to find out what pets they owned.

95 clients took part. 75 clients owned a cat (C), 30 clients owned a rabbit (R) and 15 owned both.

a) Complete the Venn diagram to show this information. [1]

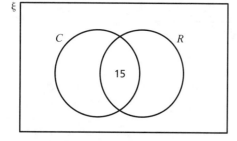

b) Write down the probability that a client owns neither a cat nor a rabbit.

Answer _____ [2]

4 A biased dice has five faces numbered 1 to 5.

The probability the dice lands on a 5 is 0.18

The dice is rolled twice and the score recorded.

a) Complete the probability tree diagram. [2]
5' means 'not 5'.

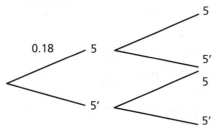

b) Calculate the probability that the dice lands on a 5 on only one of the two rolls.

Answer _____ [3]

Total Marks _____ / 17

Number Patterns and Sequences & Terms and Rules

1 Here are the first three terms in a sequence of numbers:

8, 5, 2, __, __

a) Write down the next two terms in the sequence.

Answer _____ [2]

b) Work out the expression for the nth term of the sequence.

Answer _____ [2]

c) Jennifer thinks that −15 is a number in this sequence.

Is Jennifer correct? Explain your answer.

_____ [2]

2 The following numbers form a geometric sequence.

3, 6, 12, 24, __, __

a) Write down the next two terms in the sequence.

Answer _____ [1]

b) Is 191 a term in this sequence?
Explain your answer.

_____ [2]

3 Work out the next term in this cubic sequence.

3, 17, 55, 129 …

Answer _____ [1]

4 Write down the first five terms in the sequence $3^n + 2$

Answer _____ [2]

Total Marks _____ / 12

Transformations

1

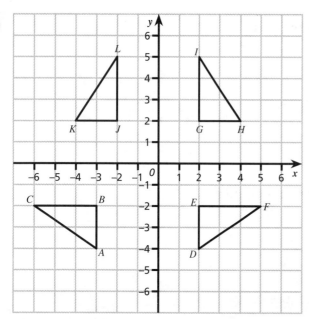

Describe the single transformation that maps:

a) Triangle *ABC* onto triangle *DEF*.

_____

_____ [2]

b) Triangle *DEF* onto triangle *GHI*.

_____

_____ [3]

c) Triangle *DEF* onto triangle *JKL*.

_____

_____ [2]

2 On the grid below plot the points: *A*(2, 1), *B*(4, 1) and *C*(3, 5). Join the points together. Using construction lines, enlarge triangle *ABC* by scale factor 2, centre of enlargement (0, 0), to form triangle *DEF*.

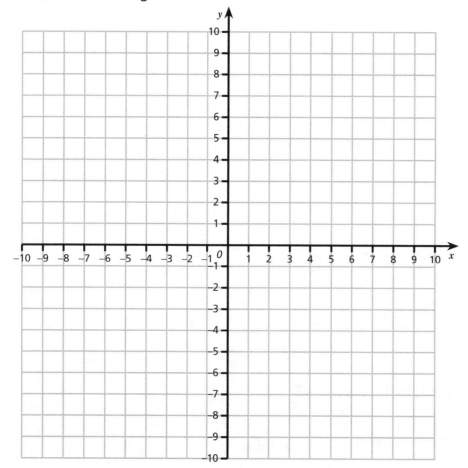

[3]

Total Marks _____ / 10

Constructions

1 Using a pair of compasses and a ruler, mark two points, C and D, that are 5cm apart.

Draw the locus of points that are equidistant from C and D.

[2]

2 Draw any triangle ABC.

Construct the bisectors of each angle using a pair of compasses and a ruler.

[3]

3 A pyramid has a rectangular base 3cm by 4cm and a height of 5cm.

Draw an accurate plan view of the pyramid.

[2]

Total Marks _____ / 7

Linear Graphs

1 Work out the equation of the line shown.

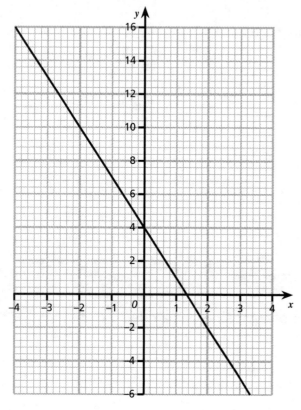

Answer _____ [3]

2 A line has the equation $5y + 2x = 7$

Work out the gradient of the line.

Answer _____ [1]

3 Work out the equation of the line that goes through points (1, 5) and (6, 15).

Answer _____ [3]

4 A graph crosses the y-axis at the point (0, 5) and the x-axis at the point (5, 0).

Write down the equation of the line in the form $ax + by + c = 0$, where a, b and c are integers.

Answer _____ [2]

Total Marks _____ / 9

Graphs of Quadratic Functions

1 A graph has the equation $y = 2x^2 - 7$

a) Complete the table below.

x	-2	-1	0	1	2
y					

[1]

b) Plot the graph of the equation on the axes below.

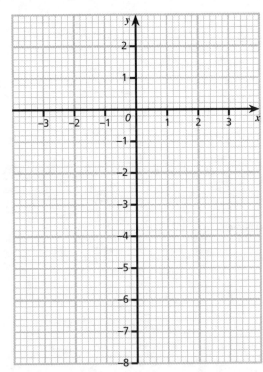

[2]

2 The graph of $y = f(x)$ is shown below.

The maximum is the point (–0.5, 0.5)

The minimum is the point (0.75, –1.5)

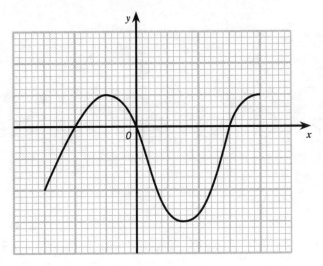

Write down the coordinates of the maximum and minimum points for the graphs of the following equations:

a) $f(x + 2)$

Maximum: _____

Minimum: _____ [2]

b) $-f(x)$

Maximum: _____

Minimum: _____ [2]

Total Marks _____ / 7

Powers, Roots and Indices

1 $\sqrt{8} + 2\sqrt{2} = k\sqrt{2}$

Work out the value of k.

Answer _____ [2]

2   Simplify $\left(2x^2 y\right)^3$

Answer _____ [2]

3   Work out the value of $\left(\frac{9}{64}\right)^{\frac{1}{2}}$

Answer _____ [1]

4   Rationalise $\dfrac{\sqrt{3} - 1}{\sqrt{3}}$

Answer _____ [2]

5   Work out the area of the rectangle. Give your answer in the form $a - b\sqrt{3}$.

$\sqrt{3}$

$\sqrt{3} - 2$

Answer _____ [2]

Total Marks _____ / 9

Area and Volume 1 & 2

1 The diagram below is the cross-section of a swimming pool.

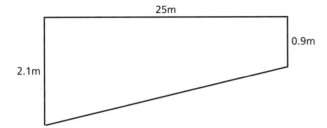

25m

0.9m

2.1m

The swimming pool is 10m wide. The pool fills at a rate of 0.2m³ per second.

How many hours does it take to fill the pool completely?

Give your answer to 3 significant figures.

Answer _____ [4]

2 The ratio of the radius to the height of a cylinder is 1 : 3
The volume of the cylinder is 275πcm³.

Calculate the value of the radius. Give your answer to 3 significant figures.

Answer _____ [4]

Total Marks _____ / 8

Uses of Graphs

1 x is inversely proportional to y.

Sketch the graph of this relationship.

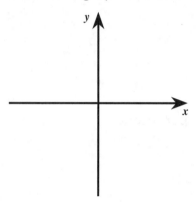

[1]

2   A line has the equation $2y = 3x + 5$

Work out the equation of the line that is perpendicular to it and goes through point (3, 6).

Answer _____ [4]

3   The formula $C = 3M + 2$ represents how the cost of a phone call is calculated by a telephone company, where C is the cost in pence and M is the number of minutes.

Write down the gradient of the line and use it to describe the rate of change.

_____ [2]

Total Marks _____ / 7

Other Graphs 1

1 The graph below shows the journey of a car.

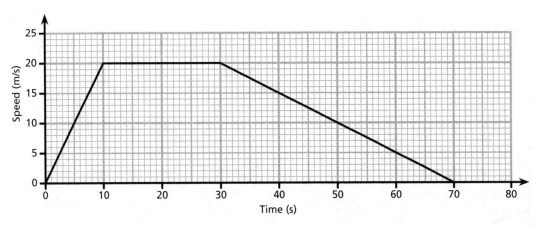

a) Describe the journey of the car.

_____

_____

_____ [3]

b) Calculate the distance covered by the car.

Answer _____ [3]

2 Sketch the graph of $y = x^3 - 1$ and label the y-intercept.

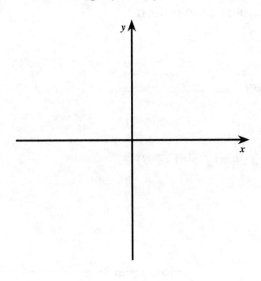

[2]

Total Marks _____ / 8

1 A circle has the equation $x^2 + y^2 = 45$

Write down the value for the length of its radius.
Give your answer in the form $a\sqrt{b}$.

Answer _____ [2]

2 A ball is thrown up in the air. The speed of the ball over the first three seconds is shown in the graph below.

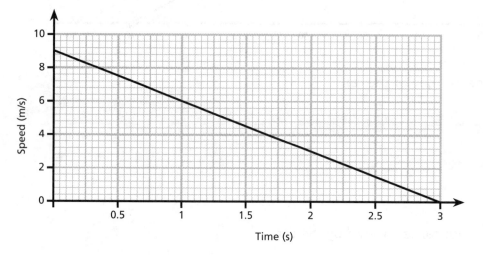

a) Write down the initial speed of the ball.

Answer _____ [1]

b) Explain what happens to the ball at three seconds.

_____ [1]

c) Work out the distance travelled by the ball in the three-second period.

Answer _____ [2]

3 A circle has the equation $x^2 + y^2 = 58$

Work out the equation of the tangent that meets the circle at point (3, 7).

Answer _____ [4]

Total Marks _____ / 10

Inequalities

1 If $-6 \leqslant d \leqslant 2$ and $-5 \leqslant e \leqslant 5$, work out: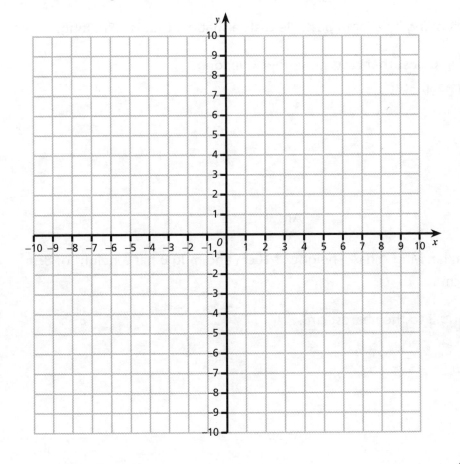

   a) The largest possible value of $d \times e$.

                                       Answer _____ [1]

   b) The smallest possible value of $d \times e$.

                                         Answer _____ [1]

2 Work out the values of y that satisfy these two inequalities: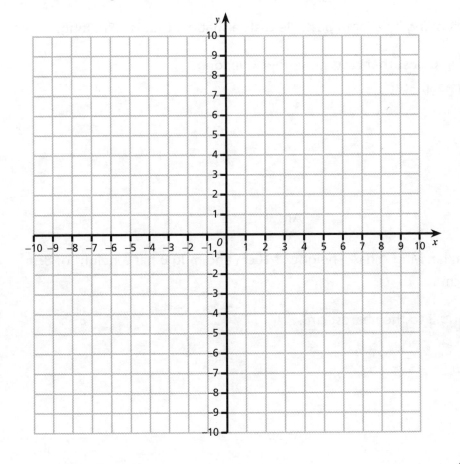

   $2y > -12$ and $3y + 4 \leqslant 19$

                                                   [3]

3 On the grid, plot the graphs of $x = 0$, $x + y = 10$ and $y = x$.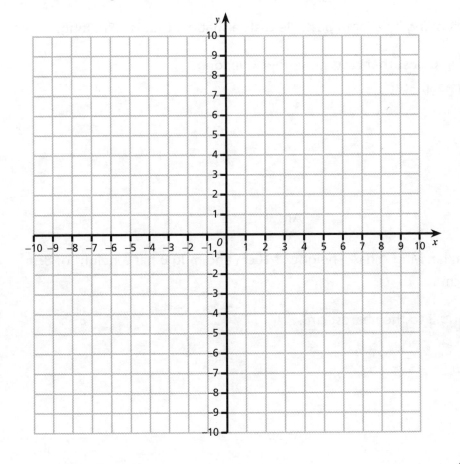

   Shade the region that represents the inequalities $x \geqslant 0$, $x + y \leqslant 10$ and $y \geqslant x$.

                                                   [4]

Total Marks _____ / 9

Congruence and Geometrical Problems

1 Here are four triangles, A, B, C and D.

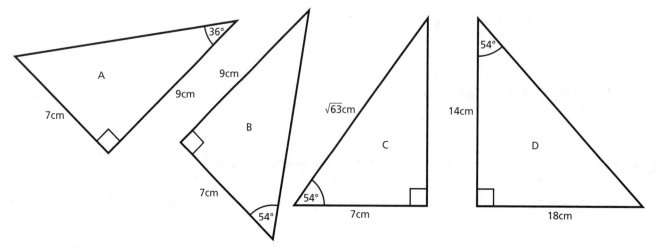

a) Which two triangles are congruent? Give a reason for your answer.

Answer _____ [2]

b) Which triangles are similar to triangle A?

Answer _____ [1]

2 A man who is 1.6m tall is standing by a lamp post. He casts a shadow that is 2.8m long.

Work out the height of a lamp post that casts a shadow 38m long.
Give your answer to 1 decimal place.

Answer _____ [2]

3 Two mugs, A and B, are similar. Mug A has a height of 10cm and mug B has a height of 8cm.
Mug A has a volume of 36cm³.

Work out the volume of mug B to the nearest cm³.

Answer _____ [3]

Total Marks _____ / 8

Right-Angled Triangles

1 The diagonal of a square has a length of 16cm.

Calculate the square's side length to 2 decimal places.

Answer _____ [3]

2 A cuboid box has a length of 13cm, a width of 12cm and a height of 5cm.

Is it possible to fit a pencil of length 18cm into the box? You **must** show your working.

[3]

3 From the top of a lookout tower of height 20m, a lifeguard sees two boats.

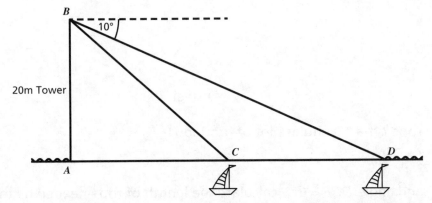

If the angles of depression of the two boats are 10° and 29°, calculate the distance between the two boats. Give your answer to the nearest metre.

Answer _____ [3]

Total Marks _____ / 9

Sine and Cosine Rules

1 Two bees, Buzz and Hum, leave a hive simultaneously.
Buzz flies 32m due south and Hum flies 19m on a bearing of 133°.

a) How far apart are the two bees? Give your answer to 2 decimal places.

Answer _____ [3]

b) What is the area of the region enclosed by the hive, Hum and Buzz?
Give your answer to the nearest square metre.

Answer _____ [2]

c) Hum flies at 40 centimetres per second. How long will it take Hum to reach Buzz?

Answer _____ [2]

2 The forensic police rope off a triangular plot of ground DEF.
DE = 30.4m and DF = 32m.

If angle FDE = 52° and angle DEF = 67°, calculate the length of rope needed by the
police to cordon off the area. Give your answer to 1 decimal place.

Answer _____ [3]

Total Marks _____ / 10

Statistics 1

1 The scatter diagram below shows the total time spent studying and the mark obtained in a mathematics test for 11 students.

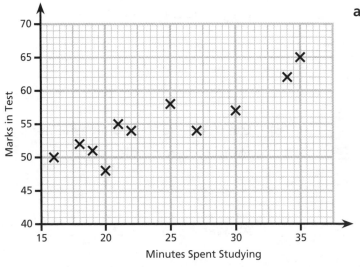

a) Another student took the test. They spent 15 minutes studying and scored 46 marks.

Add this student to the scatter diagram. [1]

b) Write down the type of correlation shown by the graph.

Answer _____ [1]

c) Explain what this correlation might mean.

_____ [1]

d) Draw a line of best fit. [1]

e) Use your line of best fit to estimate the time spent studying by a student who scored a mark of 60 in the test.

Answer _____ [1]

2 Rebecca calculates the mean time taken for her class to run 100m.
She calculates a mean of 15.2 seconds for her 23 classmates, but realises she has accidentally used a time of 21.3 instead of 12.3

Correct the mean time taken.

Answer _____ [3]

Total Marks _____ / 8

Statistics 2

1 The cumulative frequency curve below shows the length of 36 leaves taken from a tree.

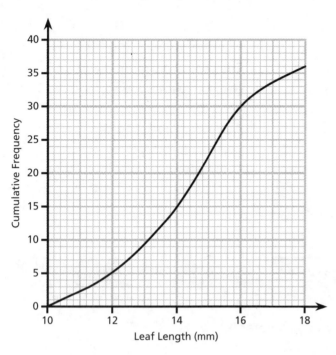

a) Use the curve to estimate the median leaf length.

Answer _____ [1]

b) Use the curve to find an estimate for the interquartile range.

Answer _____ [1]

c) Draw a box plot to represent this data.

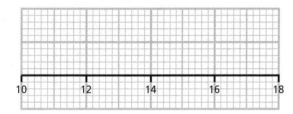

[3]

d) The box plot below shows the data from another tree.
Compare the length of leaves on the two trees.

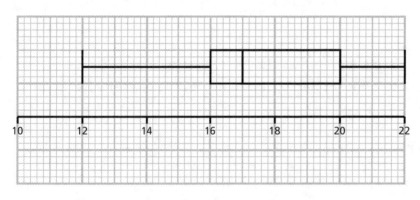

_____

_____ [2]

Total Marks _____ / 7

Measures, Accuracy and Finance

1 Estimate the value of $(8.64 \times 0.864) \div (8.4 - 8)$ 🖩

Answer _____ [2]

2 Members of the Crawshaw family enjoy camping holidays and often take their car to France by ferry. The ferry prices for a two-way trip are: cars £46, plus £12.50 per adult and £8.70 per child.

The campsite has a daily charge of €12.50 for cars, €10 per adult and €5 per child. The rate of exchange is €1 = £0.78

Calculate the total cost in pounds for the two parents and two children in the Crawshaw family to:

a) Make a single ferry crossing.

Answer _____ [2]

b) Stay for three days at the French campsite.

Answer _____ [2]

3 Find the largest and smallest possible areas of a rectangle that measures 6cm by 7cm, where each length is correct to the nearest centimetre.

Largest area = _____ cm² [1]

Smallest area = _____ cm² [1]

Total Marks _____ / 8

Solving Quadratic Equations

1 **a)** Write $x^2 + 2x - 5$ in the form $(x + p)^2 + q$, where p and q are integers.

Answer _____ [2]

b) Use your answer to part **a)** to solve the equation $x^2 + 2x - 5 = 0$
Give your answers in surd form.

Answer _____ [2]

c) Write down the coordinates of the turning point of $x^2 + 2x - 5$.

Answer _____ [1]

2 A square has side length $x + 3$ cm. The numerical value for the area of the square is equal to that of the perimeter.

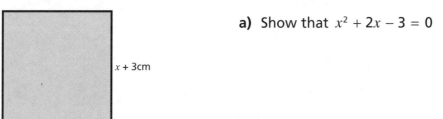
$x + 3$cm

a) Show that $x^2 + 2x - 3 = 0$

[4]

b) Work out the side length of the square.

Answer _____ [2]

Total Marks _____ / 11

Simultaneous Equations and Functions

1 Solve the simultaneous equations:

$x^2 + y^2 = 10$

$y = x + 2$

Answer _____ [4]

2 $f(x) = 3x + 5$

$g(x) = x^2 - 2x$

a) Work out the value of $f(3)$

Answer _____ [1]

b) Work out $gf(x)$

Answer _____ [2]

c) Solve $gf(x) = 0$

Answer _____ [2]

Total Marks _____ / 9

Algebraic Proof

1 Show that $(2a - 1)^2 - (2b - 1)^2 = 4(a - b)(a + b - 1)$

[4]

2 Prove that the difference between the squares of any two consecutive odd numbers is a multiple of 8.

[4]

3 Show that $0.888\,888\,88\ldots$ can be written as $\frac{8}{9}$.

[4]

Total Marks _____ / 12

Circles

1 Work out the size of angle *BAE*. Give a reason for your answer.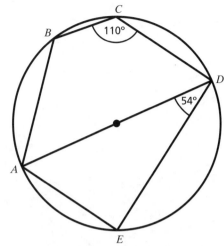

Answer _____ [3]

2 Work out the value of *d*.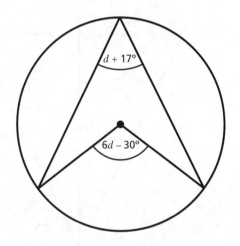

Answer _____ [2]

3 Calculate the size of angles *a*, *b*, and *c*.

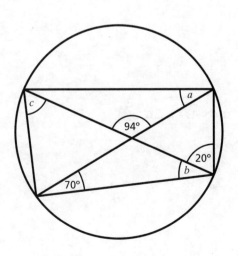

a = _____ [1]

b = _____ [1]

c = _____ [1]

Total Marks _____ / 8

Vectors

1 Here are five vectors:

$\vec{KL} = 4p + 8q$, $\vec{MN} = 8p + 16q$, $\vec{OP} = -4p + 8q$, $\vec{QR} = 12p - 24q$
and $\vec{ST} = 12p + 24q$

Which vectors are parallel?

Answer _____ [3]

2 $ABCD$ is a kite. The diagonals BD and AC intersect at E.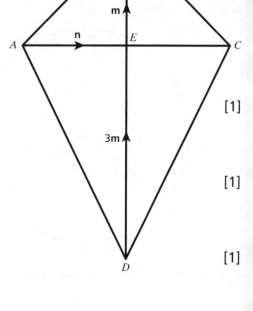

$\vec{AE} = n$, $\vec{DE} = 3m$, $\vec{EB} = m$

Work out the vector expressions for:

a) $\vec{DB}$

Answer _____ [1]

b) $\vec{CA}$

Answer _____ [1]

c) $\vec{DA}$

Answer _____ [1]

d) $\vec{AB}$

Answer _____ [1]

3 $ABCD$ is a quadrilateral where $\vec{AB} = a$, $\vec{BC} = b$, $\vec{CD} = c$ and $\vec{AD} = 2b$

Name the type of quadrilateral.

Answer _____ [1]

Total Marks _____ / 8

Collins

GCSE
Mathematics
Paper 1 Higher Tier (Non-Calculator)

Time: 1 hour 30 minutes

You must have:

- Ruler graduated in centimetres and millimetres, protractor, pair of compasses, pen, HB pencil, eraser.

You may not use a calculator

Instructions

- Use **black** ink or black ball-point pen.
- Answer **all** questions.
- Answer the questions in the spaces provided – *there may be more space than you need.*
- **Calculators may not be used.**
- Diagrams are NOT accurately drawn, unless otherwise indicated.
- You must **show all your working out**.

Information

- The total mark for this paper is 80.
- The marks for **each** question are shown in brackets
  - *use this as a guide as to how much time to spend on each question.*
- Read each question carefully before you start to answer it.
- Keep an eye on the time.
- Try to answer every question.
- Check your answers if you have time at the end.

Name: ....

Practice Exam Paper 1

1   What is one million minus one?

(Total for Question 1 is 1 mark)

2   Sandra says:

"If I double a prime number and then add three, I will always get a prime number for the answer."

Write down two numbers for which this is true and two numbers for which it is not true.
Do not use the numbers 2 or 3.

True: .... and ....

Not true: .... and ....

(Total for Question 2 is 2 marks)

3   Write fifty million in standard form.

(Total for Question 3 is 1 mark)

4   Expand and simplify $(y + 4)(y - 7)$

(Total for Question 4 is 2 marks)

5 Given that $a = \frac{1}{2}$, $b = \frac{1}{4}$ and $c = -\frac{1}{8}$, work out the value of $a^2b - c$

(Total for Question 5 is 2 marks)

6 The force (F) between two magnets is inversely proportional to the square of the distance (d) between them.

Work out the constant of proportionality if $F = 7$ when $d = 4$.

(Total for Question 6 is 2 marks)

7 The mean weight of five men is 85 kg.
The weights of four of the men are 88 kg, 91 kg, 77 kg and 94 kg.

Calculate the weight of the fifth man.

(Total for Question 7 is 3 marks)

8

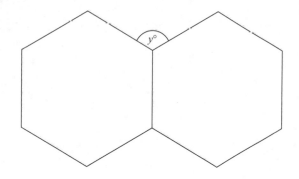

The diagram shows two regular hexagons.

Work out the size of angle y.

(Total for Question 8 is 4 marks)

9   A children's zoo has 18 rabbits.
   One full sack of rabbit food will last the zoo for 20 days.
   Another 12 rabbits are added.

   Work out how many days one full sack of rabbit food will now last.

(Total for Question 9 is 4 marks)

10   The base of a triangle ABC has been drawn for you. It is 16 cm in length.

The area of triangle ABC is always 24 cm^2.

Vertex C is a moving point.

Draw the locus of the vertex C.

A _____ B

(Total for Question 10 is 2 marks)

11   Henry and Harry are two brothers.

Henry is 9 years old and Harry is 11 years old.

£14 000 is to be shared between them in the ratio of their ages.

Work out the difference in Harry's share, if the money is shared between the brothers in 15 years' time compared to now.

Give your answer as an approximate percentage change.

(Total for Question 11 is 5 marks)

12   A bag contains eight marbles.
     Two marbles are yellow and six marbles are blue.
     A marble is taken out of the bag and not replaced.
     A second marble is taken.

     Work out the probability of taking:

     (a) Two yellow marbles.

(3)

     (b) One yellow marble and one blue marble.

(3)

(Total for Question 12 is 6 marks)

13 Write down the value of:

(a) 14^0

(1)

(b) $16^{\frac{1}{2}}$

(1)

(c) $\left(\frac{1}{8}\right)^{-1}$

(1)

(Total for Question 13 is 3 marks)

14 Louise thought of a number (n). She multiplied it by 8 and then added 13.
Her answer was less than 150.

(a) Write down an inequality for this information.

(2)

(b) n is a prime number.

What is the largest value for n that Louise could be thinking of?

(1)

(Total for Question 14 is 3 marks)

15 A solid gold sphere of radius 6 cm is melted down and recast into a solid cone.

The cone also has a radius of 6 cm.

Use $\pi = 3$

> Volume of a sphere $= \frac{4}{3}\pi r^3$
>
> Volume of a cone $= \frac{1}{3}\pi r^2 h$

(a) Work out the height of the cone.

(3)

(b) The density of gold = 20 g/cm³.

Gold sells for £20 per gram.

Work out the selling price of the gold in the sphere.

(4)

(Total for Question 15 is 7 marks)

16 An 8 m ladder leans against a vertical wall.
The ladder reaches 4 m up the wall.

Work out the angle that the ladder makes with the ground.

(Total for Question 16 is 2 marks)

17 Change the recurring decimal $0.0\dot{9}$ into a fraction and simplify.

(Total for Question 17 is 3 marks)

18   **(a)** Draw the graph of $y = \frac{36}{x}$ for the positive values of x.

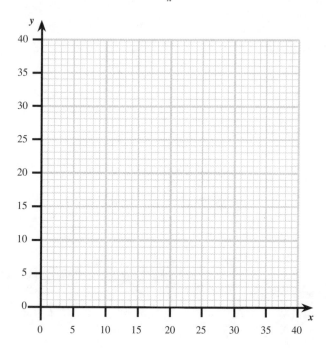

(2)

(b) Write down the name of the type of curve produced.

(1)

(c) Another line could be drawn on the axes above to solve the equation $x^2 + 3x - 36 = 0$

Work out the equation of this line.

(3)

(Total for Question 18 is 6 marks)

19

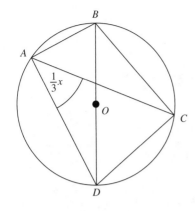

O is the centre of the circle.

Prove that:

(a) Angle $DCO = 90 - \frac{1}{3}x$

(3)

(b) Angle $BOC = 180 - \frac{2}{3}x$

(2)

(Total for Question 19 is 5 marks)

20 In the first year, a new car's value will depreciate by 15%.
Sandra paid £25 500 for a car that was one year old.

Work out what it would have cost Sandra to buy the same car when it was brand new.

(Total for Question 20 is 3 marks)

21 The area in which a discus is thrown is a sector of a circle.
The angle of the sector is 70° and the radius is 90 metres.

Taking π as $\frac{22}{7}$, work out the arc length of the sector.

(Total for Question 21 is 3 marks)

22 *ABCD* is a parallelogram.

M is the midpoint of *AB* and *N* is the midpoint of *BC*.
$\overrightarrow{DA} = \mathbf{a}$ and $\overrightarrow{DC} = \mathbf{c}$

Write down the following in terms of **a** and **c**:

(a) $\overrightarrow{DN}$

(2)

(b) $\overrightarrow{NM}$

(2)

(c) $\overrightarrow{CA}$

(2)

(d) What can be stated about *AC* and *MN*?

(2)

(Total for Question 22 is 8 marks)

23   One solution to the equation $2y^2 - 5y + m = 0$ is $y = -\frac{1}{2}$

Work out the value of m.

.

(Total for Question 23 is 3 marks)

TOTAL FOR PAPER IS 80 MARKS

Collins

GCSE
Mathematics
Paper 2 Higher Tier (Calculator)

H

Time: 1 hour 30 minutes

You must have:

- Ruler graduated in centimetres and millimetres, protractor, pair of compasses, pen, HB pencil, eraser, calculator.

Instructions

- Use **black** ink or black ball-point pen.
- Answer **all** questions.
- Answer the questions in the spaces provided – *there may be more space than you need.*
- **Calculators may be used.**
- If your calculator does not have a π button, take the value of π to be 3.142 unless the question instructs otherwise.
- Diagrams are NOT accurately drawn, unless otherwise indicated.
- You must **show all your working out**.

Information

- The total mark for this paper is 80.
- The marks for **each** question are shown in brackets
  - *use this as a guide as to how much time to spend on each question.*
- Read each question carefully before you start to answer it.
- Keep an eye on the time.
- Try to answer every question.
- Check your answers if you have time at the end.

Name: ....

Practice Exam Paper 2

Answer ALL questions.

Write your answers in the spaces provided.

You must write down all stages in your working.

1   Write the following in order of size, where a is an integer.
    Start with the smallest.

$3a^2$ $(3a)^2$ $(2.5a)^2$

(Total for Question 1 is 1 mark)

2   £180 is divided in the ratio $7 : 8$

    What is the value of the smallest share?

(Total for Question 2 is 1 mark)

3   If two 6-sided dice are thrown, what is the probability of a total score of 3?

(Total for Question 3 is 1 mark)

4   Work out $\frac{3}{7} + \frac{1}{3}$
    Give your answer as a fraction.

(Total for Question 4 is 1 mark)

5   Use your calculator to work out $(5.6 + 2.1)^2 \times 1.03$

Write down all the figures on your calculator.

(Total for Question 5 is 1 mark)

6   If $m = 5$ and $g = 6$ and each number is given to one significant figure, work out the smallest possible value of $m + g$.

(Total for Question 6 is 1 mark)

7   A uniform, cylindrical cast iron rod has a mass of 270 kg.

Its cross-section has an area of 12 cm^2.

Its density is 7.5 g/cm^3.

Calculate the volume of the rod.

(Total for Question 7 is 2 marks)

8   A regular polygon has an interior angle of 150°.

How many sides does it have?

(Total for Question 8 is 2 marks)

9   **(a)** Prove that $x^2 - 2x + 2 = 0$ has no solutions.

(3)

(b) $(x + 5)(x - 6)(x + 2)$ can be written in the form $ax^3 + bx^2 + cx + d$ where a, b, c and d are integers.

Show that $-d + c = 14(a + b)$

(4)

(Total for Question 9 is 7 marks)

10 A rectangular swimming pool has internal dimensions of 14 m wide by 24 m long and a constant depth of 1.4 m.
It is made from concrete.
The walls and base of the pool are 10 cm thick.

(a) Work out the capacity of the swimming pool in litres.

(2)

(b) Water is pumped into the pool at a rate of two litres per second.

Calculate how long it would take to fill the swimming pool to the nearest hour.

(3)

(c) Calculate the volume of concrete, in cubic metres, that would be needed to make the swimming pool.

(3)

(Total for Question 10 is 8 marks)

11 Mandeep invests £3000 in an account for six years.
The account pays 2% compound interest per annum.

How much money will Mandeep have earned in interest at the end of six years?

(Total for Question 11 is 3 marks)

12 A goat is tethered to a post by a rope that is 12 metres long.
The rope is cut in half and tied to the same post.

What fraction of the original grazing area can the goat now eat?

(Total for Question 12 is 3 marks)

13   A plane leaves Gatwick airport at 07:25 and flies due north to Hull.
It averages a speed of 560 mph and takes 25 minutes to arrive at Hull.
The plane stays at Hull for 50 minutes.
It then flies due west to Blackpool at an average speed of 420 mph.
The journey time from Hull to Blackpool is 20 minutes.

(a) At what time does the plane arrive at Blackpool?

(1)

(b) The plane returns directly from Blackpool to Gatwick.

Calculate the distance from Blackpool to Gatwick to the nearest mile.

(4)

(Total for Question 13 is 5 marks)

14   **(a)** On the axes below, draw the graphs of the equations $3y + 2x = 12$ and $y = x - 1$

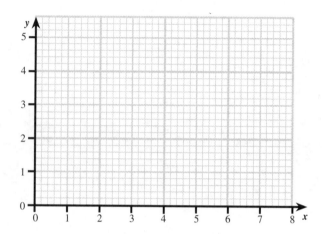

(2)

(b) Use the graphs to solve the simultaneous equations:

$3y + 2x = 12$

$y = x - 1$

(1)

(c) On the same axes, shade the region which satisfies $3y + 2x > 12$, $y < x - 1$ and $y < 2$

(3)

(Total for Question 14 is 6 marks)

15 A stone is dropped over the edge of a cliff.

The distance it falls (h) is directly proportional to the square of the time (t).

$h = 0.8$ metres when $t = 0.4$ seconds

Calculate the value of h when $t = 16$ seconds.

(Total for Question 15 is 3 marks)

16 A line has the equation $y = 4x + 6$

Write down the equation of the line that is perpendicular to it and which passes through the point $(0, 4)$.

(Total for Question 16 is 3 marks)

17   The diagram represents the ground floor of Denis and Theresa's house.

All measurements are in metres.

The kitchen is a square of side length x.

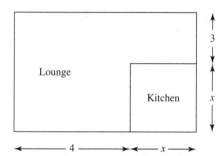

(a) Write down an expression for the perimeter of the ground floor.

(2)

(b) The perimeter of the ground floor is 24 metres.

Calculate the area of the lounge.

(3)

(c) Carpet costs £24 per square metre. In a sale it is reduced by 12%.
Denis and Theresa have put aside £625 to buy carpet for the lounge.

Is this enough money?
You must show how you got your answer.

(3)

(Total for Question 17 is 8 marks)

18 A rectangle has side lengths of $12 + \sqrt{7}$ and $12 - \sqrt{7}$ metres.

Work out:

(a) The perimeter of the rectangle.
Give your answer in its simplest form.

(2)

(b) The area of the rectangle.
Give your answer in its simplest form.

(3)

(Total for Question 18 is 5 marks)

19 A triangle has sides of lengths 7 cm, 11.3 cm and 12 cm.

Calculate:

(a) The size of the largest angle.
  Give your answer correct to 1 decimal place.

(4)

(b) The area of the triangle.
  Give your answer correct to 1 decimal place.

(4)

(Total for Question 19 is 8 marks)

20

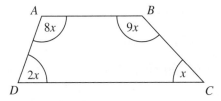

Prove that shape $ABCD$ is a trapezium.

21   Under certain conditions bacteria grow at a compound rate of 8% per hour.
      At the start there are 150 bacteria.

      Calculate the time it will take for a culture of 150 bacteria to grow to 6000 bacteria.
      Give your answer to the nearest hour.

(Total for Question 21 is 5 marks)

TOTAL FOR PAPER IS 80 MARKS

Collins

GCSE

Mathematics

Paper 3 Higher Tier (Calculator)

H

Time: 1 hour 30 minutes

You must have:

- Ruler graduated in centimetres and millimetres, protractor, pair of compasses, pen, HB pencil, eraser, calculator.

Instructions

- Use **black** ink or black ball-point pen.
- Answer **all** questions.
- Answer the questions in the spaces provided – *there may be more space than you need.*
- **Calculators may be used.**
- If your calculator does not have a π button, take the value of π to be 3.142 unless the question instructs otherwise.
- Diagrams are NOT accurately drawn, unless otherwise indicated.
- You must **show all your working out**.

Information

- The total mark for this paper is 80.
- The marks for **each** question are shown in brackets
  - *use this as a guide as to how much time to spend on each question.*
- Read each question carefully before you start to answer it.
- Keep an eye on the time.
- Try to answer every question.
- Check your answers if you have time at the end.

Name: ....

Practice Exam Paper 3

1 Barry throws a fair four-sided dice, numbered one to four, until he rolls a one.

Work out the probability that Barry throws the dice:

(a) Exactly once.

(1)

(b) Exactly three times.

(2)

(c) More than three times.

(2)

(Total for Question 1 is 5 marks)

2   Triangle ABC has a right angle at C.

Angle $BAC = 54.1°$

$AB = 10.2$ cm

Work out the length of BC.

Give your answer to 3 significant figures.

(Total for Question 2 is 3 marks)

3 The shape is made up from a rectangle and an equilateral triangle.

The length of a side of the triangle is x cm.

The height of the rectangle is 1 cm less than the length of a side of the triangle.

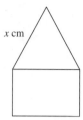

The perimeter of the shape is 38 cm.

(a) Work out the value of x.

(3)

(b) Below is a diagram of the same equilateral triangle and rectangle.

The perpendicular height of the triangle is p cm.

The length of the diagonal of the rectangle is q cm.

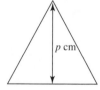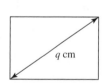

Which has a greater value, p or q?

(4)

(Total for Question 3 is 7 marks)

4   Jacob invests £8000 in an account for five years.

The account pays 2.5% compound interest per annum.

Jacob has to pay 25% tax on the interest earned each year.

The tax is taken from the account at the end of each year.

At the end of five years Jacob decides to use the money in his account to buy a new car.

The car costs £9000.

How much extra money does Jacob need to buy the new car?

(Total for Question 4 is 4 marks)

5   In the space below, use a ruler and a pair of compasses to construct an equilateral triangle of side length 5 cm. You must show all your construction lines.

(Total for Question 5 is 3 marks)

6   The table shows the duration of telephone calls (in seconds) received by two members of the customer services team at a flower pot company.

	Shortest Time (s)	Lower Quartile (s)	Median Time (s)	Upper Quartile (s)	Longest Time (s)
Bill	70	140	210	290	430
Ben	40	140	310	450	645

(a) Draw two box plots to compare both sets of data.

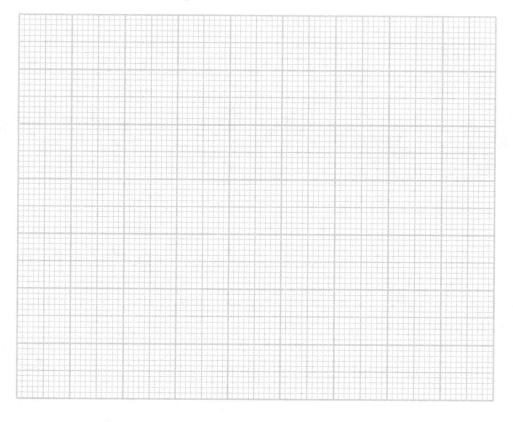

(4)

(b) Make two comments comparing the duration of the telephone calls received by Bill and Ben.

(2)

(Total for Question 6 is 6 marks)

7   *r* and *s* are positive integers.

11*r* + *s* is a multiple of 8.

Prove that 3*r* + *s* is also a multiple of 8.

(Total for Question 7 is 2 marks)

8   Five bags of sand and four bags of gravel weigh 340 kg.

Three bags of sand and five bags of gravel weigh 321 kg.

Sumira needs six bags of sand and eight bags of gravel.

Her van has a safe carrying load of 500 kg.

Sumira's friend Sam says:

"You cannot transport all the sand and gravel in one load, because it will exceed the weight that it is safe to carry."

Is Sam correct?

Explain your answer.

(Total for Question 8 is 4 marks)

9 An object has a mass of 540 g correct to 2 significant figures and a volume of 211.1 cm³ correct to 1 decimal place.

Calculate the upper and lower bounds of the density of the object.
Give your answers correct to 3 significant figures.

(Total for Question 9 is 4 marks)

10 Solve the simultaneous equations:

$2x + 3y = -3$
$3x - 2y = 28$

(Total for Question 10 is 4 marks)

11 The diagram below shows a cone with a height of 30 cm and a diameter of 15 cm.

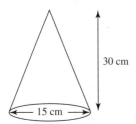

Volume of cone = $\frac{1}{3}\pi r^2 h$

(a) Work out the volume of the cone.
Give your answer in terms of π.

(2)

(b) A smaller cone of height 10 cm is removed from the top of the larger cone and frustum A is formed.

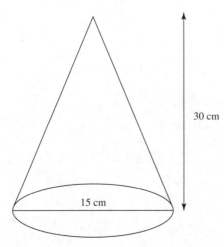
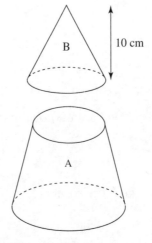

Work out the volume of frustum A.
Give your answer in terms of π.

(3)

(c) Two mathematically similar frustums have perpendicular heights of 20 cm and 30 cm.
The surface area of the smaller frustum is 450 cm².

Calculate the surface area of the larger frustum.

(2)

(d) The diagram shows a frustum.
The diameter of the base is $3d$ cm and the diameter of the top is d cm.
The perpendicular height of the frustum is h cm.

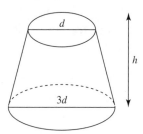

The formula for the curved surface area (s) of the frustum is:

$$s = 2\pi d\sqrt{h^2 + d^2}$$

Rearrange the formula to make h the subject.

(3)

(Total for Question 11 is 10 marks)

12 Given that $r = 3^x$ and $s = 3^y$

write down, in terms of r and s:

(a) 3^{x+y}

(1)

(b) 3^{2y}

(1)

(c) 3^{x-2}

(1)

(d) Given that $sr^2 = 9$ and $3sr^3 = 9$

work out the value of x and the value of y.

(3)

(Total for Question 12 is 6 marks)

13 Solve $\dfrac{3}{x-1} + \dfrac{2}{2x+3} = 5$

(Total for Question 13 is 4 marks)

14 Holly is designing a pendant.

The diagram below shows her design.

It is a sector of a circle, with centre O.

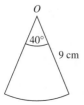

Wire is placed around the perimeter of the pendant.

Holly has 20 cm of wire.

Does she have enough wire?

You must show working to justify your answer.

(Total for Question 14 is 4 marks)

15 Work out $\dfrac{\left(4 + \sqrt{2}\right)\left(4 - \sqrt{2}\right)}{\sqrt{7}}$

Give your answer in the form $a\sqrt{b}$ where a and b are integers.

(Total for Question 15 is 4 marks)

16

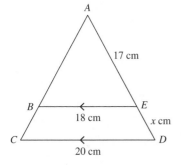

Work out the value of x.

17

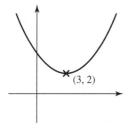

(3, 2)

The diagram above shows the graph of equation $y = f(x)$.
The coordinates of the turning point are $(3, 2)$.

(a) Write down the coordinates of the turning point of the graph with equation:

    (i) $y = f(x - 3)$

    (ii) $y = f(x) - 5$

    (iii) $y = -f(x)$

    (iv) $y = f(2x)$

(4)

(b) The graph of $y = x^2$ is translated to give the graph of $y = f(x)$.

    State $f(x)$ in terms of x.

(2)

(Total for Question 17 is 6 marks)

TOTAL FOR PAPER IS 80 MARKS

Answers

Workbook Answers

You are encouraged to show all your working out, as you may be awarded marks for method even if your final answer is wrong. Full marks can be awarded where a correct answer is given without working but, if a question asks for working, you must show it to gain full marks. If you use a correct method that is not shown in the answers below, you would still gain full credit for it.

Page 148 – Order and Value

1. a) 2×10^8 **[1]**
   b) 0.000678 **[1]**
2. $1.5 \times 10^4 = 15\,000$ **[1]**; $3.5 \times 10^3 = 3500$ **[1]**; 1.85×10^4 **[1]**
3. $\sqrt{m} = \sqrt{8100} = 90$ **[1]**; $\dfrac{3\sqrt{m}}{y} = \dfrac{270}{0.027}$ **[1]**; 1×10^4 **[1]**
4. a) 36.9 **[1]**
   b) 0.0369 **[1]**
5. 1.729 **[2]**
6. $\blacktriangle = 3$ **[1]**

Page 149 – Types of Number

1. a) $3, 7, 31$ **[1]**
   b) $3, 7$ **[1]**
   c) $12, 16, 20$ **[1]**
   d) $9, 16$ **[1]**
   e) 20 **[1]**
2. $2 \times 2 \times 19$ OR $2^2 \times 19$ **[2]**
3. $684 = 2^2 \times 3^2 \times 19$ **[1]**; $468 = 2^2 \times 3^2 \times 13$ **[1]**; HCF $= 2^2 \times 3^2 = 36$ **[1]**
4. Sum of even numbers from 2 to 1000 = 250\,500, sum of odd numbers from 1 to 999 = 250\,000 **[1]**; 500 **[1]**

> Each even number minus the odd number immediately before it equals 1. Since there are 500 even numbers from 2 to 1000, $500 \times 1 = 500$.

5. No **[1]**; because 279 and 27 are not prime numbers **[1]**

Page 150 – Basic Algebra

1. $2x + 4 = 8$ **[1]**; $2x = 4$ **[1]**; $x = 2$ **[1]**
2. -60 **[1]**
3. $3x^2 - 3xy + xy + 5y$ **[1]**; $3x^2 - 2xy + 5y$ **[1]**
4. $2x = 12$ **[1]**; $x = 6$ **[1]**
5. $\dfrac{2}{x} = 18$ **[1]**; $x = \dfrac{1}{9}$ OR 0.111 **[1]**
6. $a = 3$, $b = 3$, $c = -1$, $d = 2$, $3y(3x - y + 2x^2)$ **[2]** (1 mark for 2–3 correct terms)

Page 151 – Factorisation and Formulae

1. $x^2 + 4x - 2x - 8$ **[1]**; $x^2 + 2x - 8$ **[1]**
2. $(2x + 1)(x + 2)$ **[2]** (1 mark for each correct bracket)
3. a) 40 **[1]**
   b) $v - u = at$ **[1]**; $t = \dfrac{v - u}{a}$ **[1]**
   c) 8.75 **[1]**
4. $pr + 2p = 3r - 1$ **[1]**;
   THEN $2p + 1 = r(3 - p)$ **[1]**; $r = \dfrac{2p + 1}{3 - p}$ **[1]**

   OR $r(p - 3) = -1 - 2p$ **[1]**; $r = \dfrac{-1 - 2p}{p - 3}$ **[1]**

Page 152 – Ratio and Proportion

1. $5 : 2$ **[1]**
2. $161.7 \div 11 = 14.7$km **[1]**; $14.7 \times 13 = 191.1$km **[1]**
3. $6 : 3 : 1$, one share = £70 **[1]**; Sarah receives $6 \times £70 = £420$ **[1]**
4. $1 : 200$ **[1]**
5. a) d is proportional to t^2 so $d = kt^2$ **[1]**; $k = 5$ **[1]**
   b) $48 = 5 \times t^2$ **[1]**; $t = \sqrt{9.6} = 3.1$ seconds **[1]**

Page 153 – Variation and Compound Measures

1. $4000 \times \left(1 + \dfrac{1.6}{100}\right)^3$ **[1]**; £4195.09 **[1]**; £4195 **[1]**
2. a) Speed $= 77.8 \div 8$ **[1]**; $= 9.725$m/s **[1]**
   b) Speed $= 35010$m/h **[1]**; Speed $= 35.01$km/h **[1]**

   > Speed = Distance ÷ Time

3. Volume $= 2 \div 10.49$ **[1]**; $= 0.191$cm^3 (to 3 decimal places) OR 0.19cm^3 (to 2 decimal places) **[1]**

   > Volume = Mass ÷ Density

4. $F = \dfrac{k}{d^2} \rightarrow k = Fd^2$ **[1]**; $k = 12 \times 3^2$ **[1]**; $k = 108$ **[1]**

Page 154 – Angles and Shapes 1 & 2

1. $9x = 360°$ **[1]**; $x = 40°$ **[1]**
2. Exterior angle $(= 180° − 150°) = 30°$ **[1]**; number of sides $= 360 ÷ 30 = 12$ **[1]**
3. **a)** Correct scale drawing (see sketch below) **[1]**; distance = 62km (+/− 2km) **[1]**

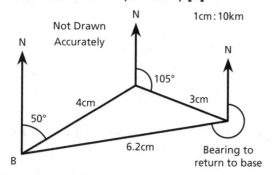

Not Drawn Accurately
1cm : 10km
N
105°
4cm
3cm
50°
6.2cm
B
Bearing to return to base

   b) Bearing 253° (+/− 2°) **[1]**

Page 155 – Fractions

1. $\frac{5}{7}$ **[1]**
2. $\frac{1}{3}$ left for the afternoon **[1]**; $\frac{2}{15}$ left for tomorrow **[1]**
3. $\frac{7}{9}$ of 81 = 63 **[1]**; $\frac{2}{7}$ of 217 = 62, so $\frac{7}{9}$ of 81 is larger **[1]**
4. 2.4 hours = 144 minutes **[1]**; $\frac{14}{144} = \frac{7}{72}$ **[1]**
5. $d = 0.272\,727...$, $100d = 27.2727...$ **[1]**; $99d = 27$ **[1]**; $d = \frac{27}{99} = \frac{3}{11}$ **[1]**
6. $\left(\frac{3}{2}\right)^p = \frac{3}{4} + \frac{3}{2} = \frac{9}{4}$ **[1]**; $p = 2$ **[1]**

Page 156 – Percentages

1. 20% increase = 180 **[1]**; 35% increase on 180 = 243 rats **[1]**
2. Convert 0.075kg to 75g **[1]**; $\frac{18}{75} \times 100 = 24\%$ **[1]**
3. $\frac{22}{100} \times c = 11$ **[1]**; $c = 50$ children **[1]**
4. $31 \times \frac{100}{62}$ OR $31 ÷ 0.62$ **[1]**; 50 plants **[1]**
5. £280 **[1]**

Page 157 – Probability 1 & 2

1. **a)** $5x + 0.5 = 1$ **[1]**; $5x = 0.5$ **[1]**; $x = 0.1$ **[1]**
   b) No **[1]**; not all outcomes are equally likely **[1]**
2. **a)** $n − 8$ **[1]**; $\frac{n − 8}{n}$ **[1]**
   b) $\frac{5}{n} \times \frac{3}{n − 1} + \frac{3}{n} \times \frac{5}{n − 1}$ **[1]**; $= \frac{30}{n(n − 1)}$ **[1]**
3. **a)** 60, 15, 5 correct on diagram (see below). **[1]**

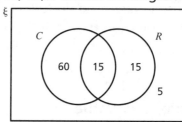

ξ
C
R
60
15
15
5

   b) $\frac{5}{95}$ **[1]**; $= \frac{1}{19}$ or 0.053 **[1]**
4. **a)** 0.82 seen **[1]**; Completely correct tree **[1]**

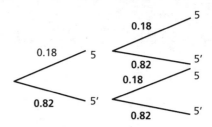

0.18
5
0.18
5
0.82
5'
0.82
5'
0.18
5
0.82
5'

   b) $0.18 \times 0.82 = 0.1476$ **[1]**; 0.1476×2 **[1]**; $= 0.2952$ **[1]**

Page 159 – Number Patterns and Sequences & Terms and Rules

1. **a)** −1, −4 **[2]**
   b) $11 − 3n$ **[2]** (1 mark for each term)
   c) No **[1]**; because n is not an integer when $11 − 3n = −15$ **[1]**
2. **a)** 48, 96 **[1]**
   b) No **[1]**; the term-to-term rule is that the numbers double each time, so 192 is a term not 191. **[1]**
3. 251 (nth term is $2n^3 + 1$) **[1]**
4. 5, 11, 29, 83, 245 **[2]** (1 mark for any three correct)

Page 160 – Transformations

1. **a)** Reflection **[1]**; in line $x = -\frac{1}{2}$ **[1]**
   b) Rotation **[1]**; anticlockwise 90° **[1]**; centre of rotation (0, 0) **[1]**
   c) Reflection **[1]**; in line $y = x$ **[1]**

2. One mark for each correct vertex of triangle *DEF* (all construction lines must be seen). **[3]**

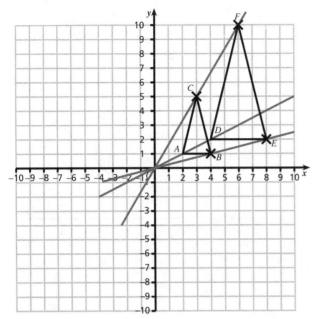

Page 161 – Constructions

1. Correct construction of the perpendicular bisector of *CD*. **[2]**

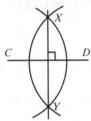

2. Correct construction of the bisectors of the three angles. **[3]**

3. Accurate rectangle 3cm by 4cm **[1]**; diagonals of rectangle drawn. **[1]**

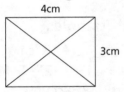

Page 162 – Linear Graphs

1. $m = -\frac{12}{4} = -3$ **[1]**; $c = 4$ **[1]**; $y = -3x + 4$ **[1]**

2. $-\frac{2}{5}$ **[1]**

3. $m = \frac{15-5}{6-1} = 2$ **[1]**; $5 = 2 \times 1 + c \rightarrow c = 3$ **[1]**;

   $y = 2x + 3$ **[1]**

4. $x + y = 5$ **[1]**; $x + y - 5 = 0$ **[1]**

Page 163 – Graphs of Quadratic Functions

1. a)

x	−2	−1	0	1	2
y	1	−5	−7	−5	1

   [1]

   b) Points plotted accurately **[1]**; joined with a smooth curve. **[1]**

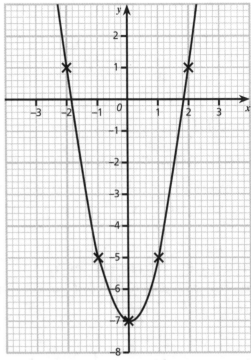

2. a) Maximum: (−2.5, 0.5) **[1]**;
      Minimum: (−1.25, −1.5) **[1]**

   b) Maximum: (0.75, 1.5) **[1]**;
      Minimum: (−0.5, −0.5) **[1]**

Page 164 – Powers, Roots and Indices

1. $\sqrt{8} = 2\sqrt{2}$ **[1]**; $k = 4$ **[1]**

2. $8x^6y^3$ **[2]** (1 mark for each correct term)

3. $\frac{3}{8}$ **[1]**

4. $\frac{\sqrt{3}-1}{\sqrt{3}} \times \frac{\sqrt{3}}{\sqrt{3}}$ **[1]**; $\frac{3-\sqrt{3}}{3}$ **[1]**

5. $\sqrt{3}(\sqrt{3} - 2)$ **[1]**; $3 - 2\sqrt{3}$ **[1]**

Page 165 – Area and Volume 1 & 2

1. Area of cross-section = $\frac{1}{2}(2.1 + 0.9) \times 25 = 37.5$

   volume = $37.5 \times 10 = 375$ **[1]**;

   $375 \div 0.2 = 1875$ seconds **[1]**; $1875 \div 3600$ **[1]**

   = 0.521 hours (to 3 significant figures) **[1]**

   3600 seconds = 1 hour

2. $h = 3r$ **[1]**; $\pi r^2(3r) = 275\pi$ **[1]**;

   $r = \sqrt[3]{\frac{275}{3}}$ **[1]**; $r = 4.51$ cm **[1]**

Page 166 – Uses of Graphs

1.

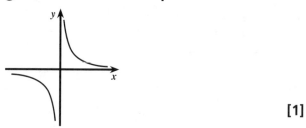

[1]

2. $m = -\dfrac{2}{3}$ [1]; $6 = -\dfrac{2}{3} \times 3 + c$ [1]; $c = 8$ [1];

$y = -\dfrac{2}{3}x + 8$ [1]

3. Gradient = 3 [1]; The cost increases by 3p per minute. [1]

Page 167 – Other Graphs 1

1. a) Accelerates for the first 10 seconds, from stationary to 20m/s [1]; drives at a constant speed for 20 seconds [1]; decelerates for the next 40 seconds to 0m/s [1]
   b) $100 + 400 + 400$ [2]; = 900m [1]

2. Correct y-intercept (0, −1) [1]; correct shape of curve. [1]

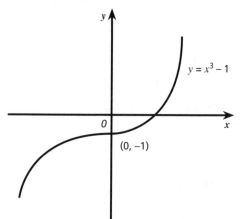

Page 168 – Other Graphs 2

1. $r = \sqrt{45}$ [1]; $3\sqrt{5}$ [1]

2. a) 9m/s   [1]
   b) It reaches its highest point and is just about to begin dropping back down   [1]
   c) $0.5 \times 3 \times 9$ [1]; = 13.5m [1]

3. Gradient of radius = $\dfrac{7}{3}$ [1]; $m = -\dfrac{3}{7}$ [1];

$7 = -\dfrac{3}{7} \times 3 + c \rightarrow c = \dfrac{58}{7}$ [1]; $y = -\dfrac{3}{7}x + \dfrac{58}{7}$

OR $7y + 3x - 58 = 0$ [1]

Page 169 – Inequalities

1. a) 30 (−6 × −5)   [1]
   b) −30 (−6 × 5)   [1]

2. $y > -6$ [1]; $y \leqslant 5$ [1]; −5, −4, −3, −2, −1, 0, 1, 2, 3, 4, 5 [1]

3. 1 mark for each correct line drawn: $x = 0$, $x + y = 10$, $y = x$ [3]; 1 mark for correctly shaded region. [1]

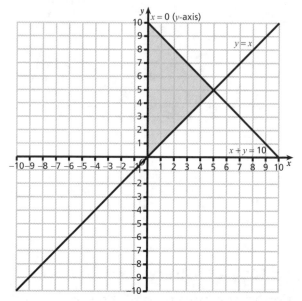

Page 170 – Congruence and Geometric Problems

1. a) A and B [1]; AAS or SAS [1]
   b) B and D   [1]

2. $\dfrac{38}{2.8} = \dfrac{\text{lamp post height}}{1.6}$ OR $(38 \div 2.8) \times 1.6$ [1]; = 21.7 metres [1]

3. Ratio of heights = $\dfrac{10}{8}$ = 1.25 [1]; Ratio of volumes = $1.25 \times 1.25 \times 1.25$ [1]; Volume of mug B = $36 \div 1.25^3 = 18\text{cm}^3$ (to the nearest cm³) [1]; OR Ratio of heights = 1 : 0.8 [1] Ratio of volumes = $1 : 0.8^3$ [1]; Volume of mug B = $36 \times 0.8^3 = 18\text{cm}^3$ (to the nearest cm³) [1]

Page 171 – Right-Angled Triangles

1. $a^2 + a^2 = 16^2$ [1]; $2a^2 = 256$ [1]; $a = 11.31$cm [1]
2. Longest diagonal of cuboid² = $13^2 + 12^2 + 5^2$ [1]; Longest diagonal of cuboid = 18.38cm [1]; Yes, 18cm pencil is shorter than the longest diagonal. [1]

3. $90° - 29° = 61°$, triangle ABC, $\tan 61° = \dfrac{AC}{20}$, AC is 36.0810m [1];

Triangle ABD, $\tan 80° = \dfrac{AD}{20}$, $AD = 113.4256$m [1]; $CD = 77$m [1]

Page 172 – Sine and Cosine Rules

1. a) Use Cosine Rule: $a^2 = b^2 + c^2 - 2bc \cos A$ **[1]**;
   $a^2 = 32^2 + 19^2 - (2 \times 32 \times 19 \times \cos 47°)$ **[1]**;
   Distance apart = 23.57m **[1]**

   b) Area $= \frac{1}{2}bc \sin A = \frac{1}{2} \times 32 \times 19 \times \sin 47°$ **[1]**;
   $= 222m^2$ **[1]**

   c) 23.57m = 2357cm **[1]**;
   Time taken $2357 \div 40 = 58.925$ seconds **[1]**

2. Use Sine Rule: $\frac{EF}{\sin 52°} = \frac{32}{\sin 67°}$ **[1]**;

   $EF = 27.4m$ **[1]**; Rope needed
   $= 32 + 30.4 + 27.4 = 89.8m$ **[1]**

Page 173 – Statistics 1

1. a) Point plotted at (15, 46) **[1]**
   b) Positive **[1]**
   c) Increase in time studying leads to
   increase in mark. **[1]**
   d) Line drawn (see diagram below) **[1]**
   e) 28–32 minutes **[1]**

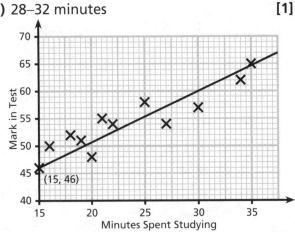

2. $15.2 \times 23 = 349.6$ **[1]**; $349.6 - 21.3 + 12.3$
   $= 340.6$ **[1]**; $340.6 \div 23 = 14.8$ **[1]**

Page 174 – Statistics 2

1. a) 14.3 – 14.7mm **[1]**
   b) 2.3 – 2.7mm **[1]**
   c) Correctly drawn lower quartile line and
   upper quartile line at either end of box
   [2]; correctly drawn median line **[1]**

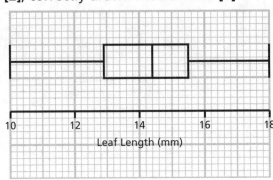

d) Tree 2 on average has longer leaves
   (higher median) **[1]**; Tree 2 has a greater
   interquartile range – more variety
   of lengths. **[1]**

Page 175 – Measures, Accuracy and Finance

1. $(10 \times 1) \div (0.5)$ **[1]**; $= 20$ **[1]** OR $(9 \times 1) \div (0.5)$ **[1]**;
   $= 18$ **[1]**

2. a) £23 + £12.50 + £8.70 **[1]**; = £44.20 **[1]**

   > Read the question carefully. The prices
   > given were for a two-way trip and the
   > question asks for costs for a single trip.

   b) $(3 \times 12.5) + 60 + 30 = 127.5$ euros **[1]**;
   127.5 euros = £99.45 **[1]**

3. Largest area $= 6.5 \times 7.5 = 48.75cm^2$ **[1]**;
   Smallest area $= 5.5 \times 6.5 = 35.75cm^2$ **[1]**

Page 176 – Solving Quadratic Equations

1. a) $(x + 1)^2 - 1 - 5 = 0$, $(x + 1)^2 - 6 = 0$ **[1]**;
   $p = 1$, $q = -6$ **[1]**
   b) $(x + 1)^2 = 6$ **[1]**; $x = -1 \pm \sqrt{6}$ **[1]**
   c) (−1, −6) **[1]**

2. a) $P = 4(x + 3)$ **[1]**; $A = (x + 3)(x + 3)$ **[1]**;
   $4x + 12 = x^2 + 6x + 9$ **[1]**;
   $x^2 + 2x - 3 = 0$ **[1]**
   b) $(x + 3)(x - 1) = 0$ **[1]**; $x = 1$,
   length = 4cm **[1]**

Page 177 – Simultaneous Equations and Functions

1. $x^2 + (x + 2)^2 = 10$ **[1]**; $2x^2 + 4x - 6 = 0$ OR
   $x^2 + 2x - 3 = 0$ **[1]**;
   $(x + 3)(x - 1)$, $x = -3$ or $x = 1$ **[1]**;
   $y = -1$ or $y = 3$ **[1]**

2. a) 14 **[1]**
   b) $(3x + 5)^2 - 2(3x + 5)$ **[1]**; $9x^2 + 24x + 15$ **[1]**
   c) $(3x + 5)(x + 1) = 0$ **[1]**; $x = -\frac{5}{3}$, $x = -1$ **[1]**

Page 178 – Algebraic Proof

1. $(2a - 1)^2 - (2b - 1)^2$
   $= 4a^2 - 4a + 1 - (4b^2 - 4b + 1)$ **[1]**;
   $= 4a^2 - 4a - 4b^2 + 4b$ **[1]**;
   $= 4(a^2 - a - b^2 + b)$ **[1]**;
   $= 4(a - b)(a + b + 1)$ **[1]**

2. $(2n + 3)^2 - (2n + 1)^2$ [1];
   $= 4n^2 + 12n + 9 - (4n^2 + 4n + 1)$ [1];
   $= 8n + 8$ [1]; $= 8(n + 1)$, a multiple of 8 [1]
3. Let $x = 0.888\,888\,88...$[1];
   $10x = 8.888\,888\,888...$ [1]; $9x = 8$ [1], $x = \frac{8}{9}$ [1]

Page 179 – Circles

1. Angle $DEA = 90°$, so angle $DAE = 36°$
   [1]; Angle $BAD = 70°$ (opposite angles of
   a cyclic quadrilateral add up to 180°) [1];
   Angle $BAE = 70° + 36° = 106°$ [1]
2. $\frac{1}{2}(6d - 30°) = d + 17°$, so $3d - 15° = d + 17°$
   OR equivalent [1]; $d = 16°$ [1]
3. $a = 16°$ [1]; $b = 16°$ [1]; $c = 74°$ [1]

Page 180 – Vectors

1. $\overrightarrow{KL}$, $\overrightarrow{MN}$, $\overrightarrow{ST}$ [2]; $\overrightarrow{OP}$, $\overrightarrow{QR}$ [1]
2. a) 4m [1]
   b) −2n [1]
   c) 3m − n [1]
   d) n + m [1]
3. Trapezium [1]

Page 181 – Practice Exam Paper 1 (Non-Calculator)

1. $1\,000\,000 - 1 = 999\,999$ [1]
2. Any two 'True' answers, e.g. 5 and 7 [1]; any
   two 'Not true' answers, e.g. 11 and 23 [1]
3. $50\,000\,000 = 5 \times 10^7$ [1]
4. $y^2 + 4y - 7y - 28$ [1]; $y^2 - 3y - 28$ [1]
5. $a^2b = \left(\frac{1}{2}\right)^2 \times \frac{1}{4} = \frac{1}{2} \times \frac{1}{2} \times \frac{1}{4} = \frac{1}{16}$ [1];
   $\frac{1}{16} + \frac{1}{8} = \frac{1}{16} + \frac{2}{16} = \frac{3}{16}$ [1]
6. $F \propto \frac{1}{d^2}$, $F = \frac{k}{d^2}$ [1]; $7 = \frac{k}{16}$, $k = 112$ [1]
7. $88 + 91 + 77 + 94 = 350$ [1]; $85 \times 5 = 425$ [1];
   weight of fifth man is $425 - 350 = 75$kg [1]
8. Interior angle of a hexagon =
   $180° - (360 \div 6)$ [1]; $= 120°$ [1];
   $y = 360° - 120° - 120°$ [1]; $y = 120°$ [1]
9. Food lasts one rabbit 18×20 days [1];
   $= 360$ days [1]; for 30 rabbits, food lasts
   360 days $\div$ 30 [1]; $= 12$ days [1]
10. Two lines drawn parallel to line AB (the
    lines should be the full page width or
    have arrows at both ends to show they are
    infinite) [1]; one line drawn 3cm above line
    AB and one line 3cm below line AB [1]
11. Now: One part is $£14\,000 \div (9 + 11) = £700$,
    Harry receives £7700 [1]; In 15 years:

£14 000 ÷ (26 + 24) = £280 [1]; Harry receives
£7280 [1]; Approximate percentage change
= (400 ÷ 8000) × 100 (or equivalent) [1];
Approximately 5% decrease (+ / − 2%) [1]

12. a) $\frac{2}{8} \times \frac{1}{7}$ [2]; $= \frac{2}{56} = \frac{1}{28}$ [1]
    b) $\frac{2}{8} \times \frac{6}{7} + \frac{6}{8} \times \frac{2}{7}$ [1]; $\frac{12}{56} + \frac{12}{56}$ [1]; $\frac{24}{56} = \frac{3}{7}$ [1]
13. a) 1 [1]
    b) 4 [1]
    c) 8 [1]
14. a) $8n + 13 < 150$ [2] (1 mark only if an
       equals symbol is used instead of
       an inequality)
    b) 17 [1]
15. a) $\frac{4\pi r^3}{3} = \frac{\pi r^2 h}{3} \rightarrow 3(4\pi r^3) = 3(\pi r^2 h)$ [1];
       $h = \frac{3(4\pi r^3)}{3\pi r^2}$, $h = 4r$ [1]; $h = 4 \times 6 = 24$cm [1]
    b) Volume = 864cm³ [1]; Mass =
       Density × Volume = $20 \times 864 = 17\,280$g [1];
       Selling price = $17\,280$g × £20 [1]; £345 600 [1]
16. $\sin \theta = \frac{\text{opp}}{\text{hyp}} = \frac{4}{8} = 0.5$ [1]; $\theta = 30°$ [1]
17. $x = 0.090\,909...$, $100x = 9.090\,909...$ [1];
    $99x = 9$, $x = \frac{9}{99}$ [1]; $\frac{1}{11}$ [1]
18. a) Correct graph of $y = \frac{36}{x}$ with points
       plotted at (1, 36), (2, 18), (3, 12), (4, 9),
       (6, 6), (9, 4), (12, 3), (18, 2) and (36, 1) [2]
       (1 mark for 6–8 correctly plotted points)

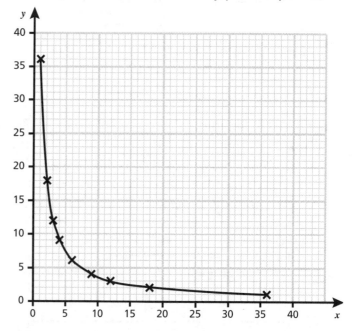

    b) Hyperbola [1]
    c) $x(x + 3) = 36$ [1]; $x + 3 = \frac{36}{x}$ [1]; Other line
       is $y = x + 3$ [1]

19. a) Correct proof showing angle $DCO = 90 - \frac{1}{3}x$, for example: Angle DAC = angle $DBC = \frac{1}{3}x$ (angles subtended from the same chord), angle $BCD = 90°$ (angle subtended from diameter BD) **[1]**; In triangle BDC, angle $BDC = 90° - \frac{1}{3}x$ **[1]**; Angle BDC = angle DCO (since triangle ODC is isosceles), therefore angle $DCO = 90° - \frac{1}{3}x$ **[1]**

   b) Correct proof showing angle $BOC = 180 - \frac{2}{3}x$, for example: Angle DBC = angle $OCB = \frac{1}{3}x$ (since triangle OBC is isosceles) **[1]**; Therefore, angle $BOC = 180° - \frac{1}{3}x - \frac{1}{3}x = 180° - \frac{2}{3}x$ **[1]**

20. 100% − 15% = 85% **[1]**; Price when new = $(£25\,500 \div 85) \times 100$ **[1]**; £30 000 **[1]**

21. Arc length = $\frac{\theta}{360} \times 2\pi r$ **[1]**; $\frac{70}{360} \times 2 \times \frac{22}{7} \times 90$ **[1]**; = 110 metres **[1]**

22. a) $\overrightarrow{DN} = \mathbf{c} + \frac{1}{2}\mathbf{a}$ **[2]**
    b) $\overrightarrow{NM} = \frac{1}{2}\mathbf{a} - \frac{1}{2}\mathbf{c}$ **[2]**
    c) $\overrightarrow{CA} = \mathbf{a} - \mathbf{c}$ **[2]**
    d) AC and MN are parallel **[1]**; $MN = \frac{1}{2}AC$ **[1]**

23. $(2y + 1)(y - 3) = 0$ **[1]**; $2y^2 - 5y - 3$ **[1]**; $m = -3$ **[1]** OR substituting in $y = -\frac{1}{2}$: $2 \times (-\frac{1}{2}) \times (-\frac{1}{2}) - 5 \times (-\frac{1}{2}) + m = 0$ **[1]**; $\frac{1}{2} + 2\frac{1}{2} + m = 0$ **[1]**; $m = -3$ **[1]**

Page 195 – Practice Exam Paper 2 (Calculator)

1. $3a^2$, $(2.5a)^2$, $(3a)^2$ **[1]**
2. £84 **[1]**
3. $\frac{1}{18}$ **[1]**
4. $\frac{16}{21}$ **[1]**
5. 61.0687 **[1]**
6. 10 **[1]**
7. $D = \frac{M}{V}$, $V = \frac{M}{D}$, $V = 270\,000 \div 7.5$ **[1]**; 36 000cm³ **[1]**
8. Exterior angle = 180° − 150° = 30° **[1]**; number of sides = 360° ÷ 30° = 12 **[1]**

9. a) $a = 1$, $b = -2$ and $c = 2$ **[1]**; $x = \frac{-b \mp \sqrt{b^2 - 4ac}}{2a} = \frac{2 \mp \sqrt{4 - (4 \times 1 \times 2)}}{(2 \times 1)}$ $= \frac{2 \mp \sqrt{-4}}{2}$ **[1]**; No solutions – impossible to square root a negative number. **[1]**

   b) $(x + 5)(x - 6)(x + 2) = x^3 + x^2 - 32x - 60$ **[1]**; $a = 1$, $b = 1$, $c = -32$, $d = -60$ **[1]**; $-d + c = 60 + -32 = 28$ **[1]**; $14(a + b) = 14 \times (1 + 1) = 28$ **[1]**

10. a) $1400\text{cm} \times 2400\text{cm} \times 140\text{cm} = 470\,400\,000\text{cm}^3$ OR $14 \times 24 \times 1.4 = 470.4\text{m}^3$ **[1]**; $470\,400\,000 \div 1000 = 470\,400$ litres OR $470.4 \times 100 = 470\,400$ litres **[1]**

    b) $470\,400 \div 2 = 235\,200$ seconds **[1]**; 235 200 seconds = 3920 minutes = 65.33 hours **[1]**; 65 hours **[1]**

    c) Outer dimensions: $14.2 \times 24.2 \times 1.5 = 515.46\text{m}^3$ **[1]**; inner dimensions: $14 \times 24 \times 1.4 = 470.4\text{m}^3$ **[1]**; volume of concrete needed = 515.46 – 470.4 = 45.06m³ **[1]**

11. $£3000 \times (1.02)^6$ **[1]**; = £3378.48 **[1]**; interest earned = £378.48 **[1]**

12. Original grazing area = $\pi r^2 = 3.142 \times 12^2 = 452.448\text{cm}^2$ **[1]**; new grazing area = $3.142 \times 6^2 = 113.112\text{cm}^2$ **[1]**; as a fraction $= \frac{113.112}{452.448} = \frac{1}{4}$ **[1]** OR Ratio of radii is 2:1 **[1]**; so, ratio of areas = 4:1 **[1]**; $\frac{1}{4}$ **[1]**

13. a) Departs Gatwick 07:25, arrives Hull 07:50, departs Hull 08:40, arrives Blackpool 09:00 **[1]**

    b) Gatwick to Hull = $560 \times \frac{25}{60} = 233.33$ miles **[1]**; Hull to Blackpool = $420 \times \frac{20}{60} = 140$ miles **[1]**; Blackpool to Gatwick (using Pythagoras' Theorem) = $\sqrt{233.33^2 + 140^2}$ **[1]**; 272 miles **[1]**

14. a) Correct graph of $3y + 2x = 12$ with points plotted at (0, 4) and (6, 0) as shown **[1]**; correct graph of $y = x - 1$ with points plotted at (0, −1), (1, 0) and (3, 2) as shown. **[1]**

    b) $x = 3$, $y = 2$ **[1]**

c) Correct shading of $3y + 2x > 12$, $y < x - 1$, $y < 2$ as shown (1 mark for each correct boundary) **[3]**

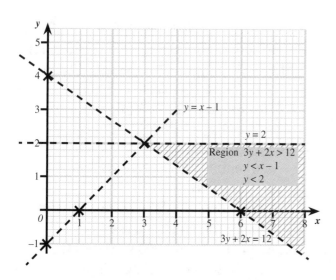

15. $h = kt^2$, $k \times 0.4 \times 0.4$ **[1]**; $k = 5$ **[1]**; $h = 5 \times 16 \times 16 = 1280$m **[1]**

16. Gradient = 4, so gradient of perpendicular line is $-\frac{1}{4}$ **[1]**; $y = mx + c$, $4 = (-\frac{1}{4} \times 0) + c$, $c = 4$ **[1]**; $y = -\frac{1}{4}x + 4$ **[1]**

17. a) $4x + 14$ (1 mark for each correct term) **[2]**
    b) $4x + 14 = 24$m, $x = 2.5$m **[1]**; Area of lounge = $(4 \times 5.5) + (2.5 \times 3)$ OR $(6.5 \times 5.5) - 2.5^2$ **[1]**; 29.5m^2 **[1]**
    c) 12% of £24 = £2.88, sale price = £21.12 **[1]**; 29.5m$^2 \times$ £21.12 = £623.04 **[1]**; Yes, it is enough money as £625 > £623.04 **[1]**

18. a) Perimeter = $2(12 + \sqrt{7}) + 2(12 - \sqrt{7})$ **[1]**; = 48m **[1]**
    b) Area = $(12 + \sqrt{7})(12 - \sqrt{7})$ **[1]**; = $144 + 12\sqrt{7} - 12\sqrt{7} - 7$ **[1]**; = 137m^2 **[1]**

19. a) $\cos A = \dfrac{(b^2 + c^2 - a^2)}{2bc}$ **[1]**;
    $\cos A = \dfrac{7^2 + 11.3^2 - 12^2}{2 \times 7 \times 11.3}$ **[1]**;
    $\cos A = 0.2066$ **[1]**; largest angle = 78.1° (to 1 d.p.) **[1]**
    b) Area = $\frac{1}{2}bc \sin A$ **[1]**;
    Area = $\frac{1}{2} \times 7 \times 11.3 \times \sin 78.1°$ **[1]**;
    Area = 38.70003cm^2 **[1]**;
    38.7cm^2 (to 1 d.p.) **[1]**

20. Written proof using allied angles, for example: Sum of interior angles in a quadrilateral = 360°, $20x = 360°$, $x = 18°$ **[1]**; Angle $A = 144°$, angle $B = 162°$, angle

$C = 18°$, angle $D = 36°$ **[1]**; If $ABCD$ is a trapezium, then AB is parallel to DC **[1]**; and angle A + angle D should equal 180° (allied angles) and angle B + angle C should = 180° **[1]**; 144° + 36° = 180° and 162° + 18° = 180° **[1]**; so $ABCD$ is a trapezium. **[1]**

21. $6000 = 150 \times (1 + 0.08)^n$ **[1]**; use of trial and improvement method shown **[1]**; $n = 20 \rightarrow$ 699.1436 (too small), $n = 50 \rightarrow$ 7035.2419 (too big), $n = 45 \rightarrow$ 4788.0674 (too small), $n = 47 \rightarrow$ 5584.8018 (too small) (or equivalent working) **[1]**; $n = 48 \rightarrow$ 6031.586 (very close) **[1]**; $n = 48$ hours **[1]**

Page 209 – Practice Exam Paper 3 (Calculator)

1. a) $\frac{1}{4}$ **[1]**
   b) $\frac{3}{4} \times \frac{3}{4} \times \frac{1}{4}$ **[1]**; $\frac{9}{64}$ **[1]**
   c) $1 - \frac{1}{4} - \frac{3}{16} - \frac{9}{64}$ **[1]**; $\frac{27}{64}$ **[1]**

2. $\sin\theta = \dfrac{\text{opp}}{\text{hyp}}$, $\sin 54.1° = \dfrac{BC}{10.2}$ **[1]**; $BC = 8.26242$ **[1]**; $BC = 8.26$ to 3 significant figures **[1]**

3. a) $3x + 2x - 2 = 38$ **[1]**; $5x = 40$ **[1]**; $x = 8$cm **[1]**
   b) $q = \sqrt{8^2 + 7^2}$ **[1]**; half base of triangle = 4cm **[1]**; $p = \sqrt{8^2 - 4^2}$ **[1]**; q is greater **[1]**

4. $0.025 \times 0.75 = 0.01875$ (75% of 2.5%) **[1]**; £8000 $\times (1 + 0.01875)^5$ = £8778.66 **[1]**; £9000 − £8778.66 **[1]**; £221.34 **[1]**

5. 1 mark for each accurately drawn side of exactly 5cm **[3]**

6. a) Correctly plotted lower and upper quartiles for Bill **[1]**; correctly plotted median line for Bill **[1]**; correctly plotted lower and upper quartiles for Ben **[1]**; correctly plotted median line for Ben **[1]**

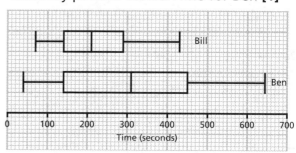

   b) On average Ben's phone calls are of longer duration **[1]**; The duration of Bill's phone calls is more consistent **[1]**

7. $(11r + s) - (3r + s) = 8r$ **[1]**; $8r$ is a multiple of 8 **[1]**

> The difference between two multiples of 8 should be a multiple of 8.

8. $5s + 4g = 340$ and $3s + 5g = 321$ **[1]**; $15s + 12g = 1020$, $15s + 25g = 1605$, $13g = 585$ (or equivalent method), $g = 45$, $15s + 540 = 1020$, $15s = 480$ (or equivalent method), $s = 32$ **[1]**; $6 \times 32 + 8 \times 45 = 552$ **[1]**; 552kg is greater than the 500kg safe limit, so Sam is correct **[1]**

9. 540g to 2 significant figures: 535 to 545 **[1]**; 211.1 to 1 decimal place: 211.05 to 211.15 **[1]**; upper bound $= 545 \div 211.05 = 2.58$ to 3 significant figures **[1]**; lower bound $= 535 \div 211.15 = 2.53$ to 3 significant figures **[1]**

10. $6x + 9y = -9$ and $6x - 4y = 56$ **[1]**; $13y = -65$ **[1]**; $y = -5$ **[1]**; $x = 6$ **[1]**

11. **a)** $\frac{1}{3} \times \pi \times 7.5^2 \times 30$ **[1]**; $\frac{1125\pi}{2}$ **[1]**
    b) 5cm seen as diameter of smaller cone **[1]**;
    $\frac{1}{3} \times \pi \times 2.5^2 \times 10 = \frac{125\pi}{6}$ **[1]**;
    $\frac{1125\pi}{2} - \frac{125\pi}{6} = \frac{1625\pi}{3}$ **[1]**

> The cones are similar and $\frac{30}{10} = 3$cm,
> so $\frac{15}{3} = 5$cm

c) Ratio of lengths $= 20 : 30 = 1 : 1.5$, 450×1.5^2 **[1]**; 1012.5cm^2 **[1]**

> Ratio of lengths is $1 : 1.5$
> Ratio of areas is $1 : 1.5^2$

d) $\frac{s}{2\pi d} = \sqrt{h^2 + d^2}$ **[1]**; $\left(\frac{s}{2\pi d}\right)^2 - d^2 = h^2$ **[1]**;
$h = \sqrt{\left(\frac{s}{2\pi d}\right)^2 - d^2}$ **[1]**

12. **a)** rs **[1]**
    b) s^2 **[1]**
    c) $\frac{r}{9}$ **[1]**
    d) $3r = 1$, $r = \frac{1}{3}$ **[1]**; $x = -1$ **[1]**; $y = 4$ **[1]**

13. $3(2x + 3) + 2(x - 1) = 5(x - 1)(2x + 3)$ **[1]**;
    $10x^2 - 3x - 22 = 0$ **[1]**;
    $x = \frac{3 \mp \sqrt{(-3)^2 - 4 \times 10 \times (-22)}}{20}$ **[1]**;
    $x = 1.64$ and $x = -1.34$ **[1]**

14. Length of arc $= \frac{40}{360} \times 2 \times \pi \times 9$ **[1]**;
    $= 6.28$cm **[1]**; perimeter $= 6.28 + 9 + 9$
    $= 24.28$cm **[1]**; 24.28 > 20, so Holly does not have enough wire **[1]**

> Find the whole perimeter.

15. $16 + 4\sqrt{2} - 4\sqrt{2} - 2$ **[1]**; $\frac{14}{\sqrt{7}}$ **[1]**; $\frac{14}{\sqrt{7}} \times \frac{\sqrt{7}}{\sqrt{7}}$ **[1]**; $2\sqrt{7}$ **[1]**

16. $\frac{20}{18} = \frac{17 + x}{17}$ **[2]**; $17 + x = \frac{170}{9}$ **[1]**;
    $x = \frac{17}{9} = 1.89$ (to 2 decimal places) **[1]**

17. **a)** **i)** (6, 2)   **[1]**
         ii) (3, −3)   **[1]**
         iii) (3, −2)   **[1]**
         iv) (1.5, 2)   **[1]**
    b) $(x - 3)^2 + 2$ **[2]** (1 mark if only the first part of the expression is given / correct, i.e. $(x - 3)^2$)

Notes